Deep Learning for Object Detection and Localization

Abdussalam Elhanashi · Sergio Saponara

Deep Learning for Object Detection and Localization

Springer

Abdussalam Elhanashi
Department of Information Engineering
University of Pisa
Pisa, Italy

Sergio Saponara
Department of Information Engineering
University of Pisa
Pisa, Italy

ISBN 978-981-95-4845-3 ISBN 978-981-95-4846-0 (eBook)
https://doi.org/10.1007/978-981-95-4846-0

This Springer imprint is published by the registered company Springer Nature Singapore Pte Ltd.
The registered company address is: 152 Beach Road, #21-01/04 Gateway East, Singapore 189721, Singapore

I am extremely thankful to my parents whose steadfast support and guidance have acted as the foundation for my journey. Your faith in me gave me continual strength and your sacrifices allowed this to happen. To my friends, thank you for being my support system. Your encouragement, patience, and shared belief in this project made all of the difference. This success is as much yours as it is mine. Also, I give my greatest appreciation to the academic professionals who read over this manuscript. Your input and critiques were key to the writing of this project, greatly increasing its quality.

Preface

Deep Learning for Object Detection and Localization is a complete guide to object detection and localization in computer vision. The book starts by covering the basics of object detection and localization, the history of methods and how deep learning has changed everything. Image processing and preprocessing techniques are covered including data annotation, synthetic data generation and how to handle imbalanced datasets so you have a solid foundation for training your model. The book goes deep into state-of-the-art object detection algorithms and localization methods and evaluation metrics to measure model performance. Practical applications across domains like autonomous driving, surveillance, medical imaging and robotics show the real-world impact. Frameworks for deep learning, optimization procedures, and deployment techniques are framed in terms to support building models. Training methods, tuning hyperparameters, and transfer learning are linked as techniques that can be taken into consideration for achieving model performance, developing models while avoiding over-fitting. Ethics, occlusion, and real-time implementation challenge are discussed, providing context to the entire landscape of object detection and localization, current and future. This book is a structured resource to understand the theory and implement object detection solutions.

Pisa, Italy
October 2025

Abdussalam Elhanashi
Sergio Saponara

Acknowledgements Thanks are due to the University of Pisa for its unwavering support and commitment to enhancing research and education in AI and deep learning, as it has contributed depth and quality to this resource on Deep Learning for Object Detection and Localization, with their collaborative efforts of expertise and resource-sharing. Thanks are also extended to the Society for Imaging Informatics in Medicine (SIIM), for its commitments to imaging research and sharing datasets and knowledge that have strongly shaped the applications to medical imaging. Their support has been essential in creating a resource that consolidates theory and practice in real world.

Competing Interest The authors have no conflicts of interest to declare that are relevant to the content of this chapter.

Contents

Acronyms

ADAS	Advanced Driver-Assistance Systems
AI	Artificial Intelligence
ANN	Artificial Neural Network
AR	Augmented Reality
AV	Autonomous Vehicle
BiFPN	Bi-directional Feature Pyramid Network
BoF	Bag of Freebies
BoS	Bag of Specials
CNN	Convolutional Neural Network
COCO	Common Objects in Context
CPU	Central Processing Unit
CT	Computed Tomography
CV	Computer Vision
CVAT	Computer Vision Annotation Tool
DETR	Detection Transformer
DL	Deep Learning
DLA	Deep Layer Aggregation
DPM	Deformable Part Model
FCN	Fully Convolutional Network
FCOS	Fully Convolutional One-Stage Object Detector
FID	Fréchet Inception Distance
FN	False Negative
FP	False Positive
FPN	Feature Pyramid Network
FPS	Frames Per Second
GAN	Generative Adversarial Network
GDPR	General Data Protection Regulation
GIoU	Generalized Intersection over Union
GPU	Graphics Processing Unit
GRU	Gated Recurrent Unit
HOG	Histogram of Oriented Gradients
HSV	Hue, Saturation, Value
IoT	Internet of Things

IoU	Intersection over Union
IS	Inception Score
JPEG	Joint Photographic Experts Group
JSON	JavaScript Object Notation
LSTM	Long Short-Term Memory
mAP	Mean Average Precision
Mask R-CNN	Mask Region-Based Convolutional Neural Network
ML	Machine Learning
MRI	Magnetic Resonance Imaging
NAS	Neural Architecture Search
NLP	Natural Language Processing
NMS	Non-Maximum Suppression
OCR	Optical Character Recognition
ONNX	Open Neural Network Exchange
PACS	Picture Archiving and Communication System
PANet	Path Aggregation Network
PET	Positron Emission Tomography
PNG	Portable Network Graphics
R-CNN	Region-Based Convolutional Neural Network
ReLU	Rectified Linear Unit
ResNet	Residual Network
RGB	Red, Green, Blue
RNN	Recurrent Neural Network
RoI	Region of Interest
RPN	Region Proposal Network
SGD	Stochastic Gradient Descent
SIIM	Society for Imaging Informatics in Medicine
SMOTE	Synthetic Minority Over-sampling Technique
SPP	Spatial Pyramid Pooling
SSD	Single Shot Multibox Detector
SVM	Support Vector Machine
TF	TensorFlow
TFLite	TensorFlow Lite
TIFF	Tagged Image File Format
TPU	Tensor Processing Unit
VAE	Variational Autoencoder
VIA	VGG Image Annotator
VOC	Visual Object Classes
VR	Virtual Reality
WebP	Web Picture Format
XAI	Explainable Artificial Intelligence
YOLO	You Only Look Once
YUV	Luminance and Chrominance

Chapter 1
Introduction to Object Detection and Localization

Abstract This chapter presents a thorough overview of object detection and localization, essential tasks in computer vision that allow machines to detect and localize objects in images or video frames. It discusses the historic development of these techniques, from the early days of handcrafted feature-based methods such as Haar cascades and HOG to the game-changing rise of deep learning. This chapter describes the role of a Convolutional Neural Network (CNN) in automatically extracting features and supporting end-to-end learning, making models the foundation of modern approaches. Furthermore, this chapter provides an extensive taxonomy of recent architectures, including two-stage detectors (e.g., Faster R-CNN) that are optimized for high accuracy, single-stage detectors (e.g., YOLO, SSD) that maximize real-time performance, and emerging anchor-free and transformer-based approaches (e.g., DECTER). The discussion also highlights several key challenges in the area, including scale variability, occlusions, real-time inference requirements, and the requisite for large annotated datasets. Additionally, the chapter demonstrates the significance of object detection for many real-world applications, such as autonomous driving, surveillance, and medical imaging. Lastly, it outlines the main differences between the related tasks of object detection–which detects objects within individual frames–and object tracking, which preserves identity across a video sequence. This introductory picture is necessary to frame up in the forthcoming chapters where we consider in depth methodologies, applications, and future directions.

1.1 Overview

Object detection and localization are essential tasks in computer vision, which allow machines to recognize and localize objects in an image or video frame, and they are essential for some applications of computer vision, including but not limited to autonomous driving, surveillance, medical imaging, robotics, and augmented reality [1]. Earlier, these problems were addressed with strictly hand-crafted features using classical machine learning algorithms such as Haar cascades, Histogram of Oriented Gradients (HOG), and Support Vector Machines (SVM), which could work

A. Elhanashi and S. Saponara, *Deep Learning for Object Detection and Localization*,
https://doi.org/10.1007/978-981-95-4846-0_1

reasonably well on simple tasks, but struggled to generalize to complex scenes, occlusion, lighting and scale variability [2]. The rise of deep learning transformed this area since deep learning offers automated feature extraction and end-to-end learning. With advances in tools based on Convolutional Neural Networks (CNN) which are usually viewed as the backbone of modern object detection systems [3], it has become possible to design detectors that are both accurate and generalizable. Modern architectures for facilitating object detection can generally be categorized as either two-stage detectors, single-stage detectors, anchor-free, or transformer architecture based detectors. Two-stage detectors (e.g., Region-based CNNs (R-CNN), Fast R-CNN, Faster R-CNN) typically develop region proposals first before classifying and refining those proposed regions. This architecture provides high accuracy; however, it does have computational drawbacks i.e., it is not as computationally efficient [2]. Single-stage detectors like YOLO or SSD predict bounding boxes and respective class probabilities on a single-stage process, allowing good real-time performance and applications such as autonomous driving, although sacrificing some accuracy [4]. Recent works under this classificatory do away with anchors and are considered anchor-free detectors, such as CenterNet and FCOS [5]. Transformer-based models, including DETR, employ self-attention along with state-of-the-art accuracies and give a new perspective to the task of object detection [6]. Several barriers in object detection and localization are scale-change, occlusions, time-performance, and number of data. An object is represented in images at different scales, posing a difficulty to correct detection for small-scale and large-scale objects alike, a difficulty in multi-scale training and design of FPNs [4]. Detecting occluded and overlapping objects becomes a complicated task, requiring instance segmentation and attention mechanisms [7]. Real-time performance holds paramount importance in numerous applications, and it is through various optimizations such as model pruning, quantization, and efficient architectures MobileNets and EfficientNets, that faster inference times are possible without any significant compromise on accuracy [9]. A major problem lies in limited annotated data availability for training deep learning models, which is being tackled through semi-supervised learning, data augmentation, and transfer learning [9]. Hence, performance evaluation of object detection models uses metrics such as precision, recall, IoU, and mAP to determine and compare performances [1]. Further, benchmark datasets like COCO, Pascal VOC, and ImageNet provide a common platform to evaluate approaches and contribute to the COCO Detection Challenge and ILSVRC to pose challenges that push innovation [4]. Object detection thus finds applications in innumerable domains such as autonomous vehicles, surveillance, healthcare, retail, and agriculture. Real-time object detection in autonomous vehicles detects pedestrians, vehicles, and obstacles to ensure safety through perception systems [3]. As are considered laws, privacy prescriptive measures are sometimes implemented for surveillance purposes mainly activity monitoring, intruder detection, or the very behavior of an anomaly [2]. In medical imaging, the object detector has to appear for diseases, locate tumors, or analyze anatomical structures, with accuracy and interpretability being the utmost important [7]. In retail and e-commerce, object detection is an integral part of supply management, shelf monitoring, customer behavior analysis, etc., and it is also used in AR that permits virtual try-ons [4]. In agriculture, object detection facilitates crop monitoring, pest detection, and assessing plant health in precision farming, meanwhile, where AI-assisted camera drones are so very much used in the field [9]. Major concerned areas for future object detection studies may include self-supervised learning, explainability, edge computing, and ethical concerns. Self-supervised learning tries to reduce the dependence on labeled data by using unlabeled data for a pre-training task and holds much promise in object detection when fewer data are available [10]. Since standardized models become of higher and complexities, the decision-making process becomes transparent to the point that has given rise to research in explainable AI, for object detection systems be reliable [6]. Object detection model to be deployed on edge devices such as smartphones and IOT devices need a lightweight architecture for real-time deployment in resource limited settings [8]. The commodification of object-detection systems has brought to light the questions about bias, privacy and misuse that we have been grappling with as we seek to implement AI fairly and responsibly [5]. This Book What is the deep learning algorithm you want to learn (20-words about) This book is a comprehensive introduction and up-to-date reference to the concepts, methodologies and latest developments. Deep Learning for Object Detection and Localization: Become an expert in Computer Vision, Machine Learning and AI with the latest techniques explained clearly Deep learning appears to mainly BE HYPE. As deep learning technologies continue to evolve rapidly, object detection and localization will drive the most cutting edge-applications across many sectors and essentially define AI as it is used in computer vision [1]. Figure 1.1 shows Example of object detection in an urban street scene.

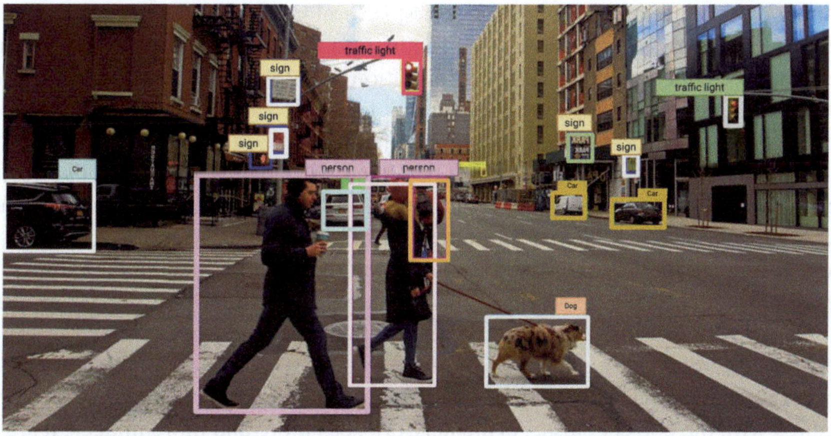

Fig. 1.1 Example of object detection in an urban street scene. The model identifies and labels multiple objects including persons, cars, traffic lights, signs, and a dog, each enclosed within bounding boxes of different colors

1.2 Understanding the Basics

Object detection-the problem of finding objects in an image, and Localizing these detected objects—i.e. to pin point their exact position on the image are essential tasks in Computer Vision. Unlike the regular image classification task where you are simply told what, recognition includes determining where the object is in the image to localize it. Object detection and localization is in a greater challenge, because the model has to deduce not only what object exists but also where it is in the image; and thus it gives both of class labels and spatial coordinates.

Object Detection and Localization: Some of the Key Factors. The two dominant bounderies in object detection and localization are.

1.2.1 Object Identification: Recognizing the Presence of Objects

Object recognition or classification is a precursor to object detection. This is the phase where the neural network learns to recognize the existence of objects present within an image. This layer deals mainly with learning abstract features or patterns within the image that differentiate between objects. This is often done with Convolutional Neural Networks (CNNs), a type of deep learning model that works particularly well with image data.

CNNs are also organized in a hierarchical manner and each layer learns a more abstract representation of the input The first several layers of the CNN are learned to identify low-level features, such as edges, texture, and color changes. Such features are basic blocks of the image. As the information flows deeper into this network, these layers combine these low-level abstractions to create even more complex patterns (i.e.., shapes, parts of objects, then whole objects).

The early layers might for instance detect edges and simple shapes, like rectangles and circles if we're detecting cars, e.g. As the data is fed into deeper layers, eventually the network begins to identify details such as headlights, wheels or car body structures. By the time data flow to deeper layers of architecture, it knows what are factors which make up a car in different perspective or lightings or occlusions.

The network is trained on a large labeled set with ground truth annotations. These labels are the class (e.g., car, dog, human) and in some cases the position of the object on an image. The network's objective during training is to associate low level features with the correct class. As the network sees more and more images, it slowly tweaks its internal settings (weights) so that it can start to categorize objects in other images.

1.2.2 *Localization: Determining the Position of the Object*

Once the network knows what is in the image, there needs to be another process to find where it is—this is known as "localization" or determining the exact position of one/between multiple objects in an image. Usually the localization is carried out by predicting a bounding box of the object. A bounding box is the rectangle that surrounds an object, and is described by four numbers: x—the coordinate at the top left of box, y—the coordinate at the top left of box, x'—bottom right x coordinates of the face rectangle, y'—bottom right y-coordinates on the face rectangle. These coordinates define the location of the object in space on the image.

The network, using learned features, estimates the coordinates of this bounding box for every object in the image. This requires the model to be sensitive not only to the properties that describe what the object is, but how those properties are located in space within an image. The result of the network is generally divided into two main stages:

- Bounding Box Coordinates: This is the values (x, y, x', y') where sampled object appears at the location in the image. They specify the dimensions of a box that surrounds the detected object.
- Confidence Score: This is the score that determines how sure the network is about that specific object being in a predicted bounding box. The higher the confidence, the more certain is the network of a correct detection.

If the network was traiend for detecting cars in an image, it would output a set of coordinates that describe where to place the car's bounding box located and how confident (+ 50 confidence = plateau) the network is that this region contains cars.

Localization is extremely important for many applications including autonomous vehicles: it is not only enough to detect pedestrians or cars, but also you need to know where exactly the pedestrian or car lies in the road (to avoid them while driving). In medical imaging, localization is used to identify the precise location of a tumor or other abnormality in an X-ray or MRI image. Figure 1.2 illustrates bounding box prediction in object detection.

1.3 Foundation of Deep Learning

Introduction Deep learning is a subset of machine learning that has transformed the filed of artificial intelligence by allowing algorithms to learn complex patterns from big data. The roots of the field are rooted in a blend of mathematics, computer science and neuroscience that have driven this development for decades. This essay looks at some of the basic principles of deep learning—historical aspects, mathematical interpretations, main architectures and interesting applications.

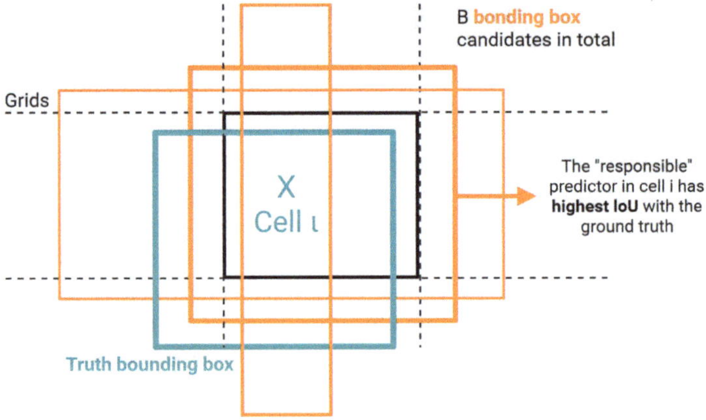

Fig. 1.2 Illustration of bounding box prediction in object detection. Each grid cell generates multiple candidate bounding boxes (orange), and the "responsible" predictor is chosen as the one with the highest Intersection over Union (IoU) relative to the ground

1.3.1 Historical Evolution of Deep Learning

Deep learning has its origins in the 1940s and 1950s, when prominent computer scientists started gaining interest in artificial neural networks (ANNs). The first model of a neural network was developed by Warren McCulloch and Walter Pitts in 1943, who based their proposal on the architecture and behavior of the human brain [21]. This was the model that set the stage for future neural network development.

The perceptron, a single-layered neural network that is used for binary classification was formulated in the 1950s by Frank Rosenblatt [24]. Despite its limitations, the perceptron was a major development in machine learning. But the field suffered a blow in the 1960s when Marvin Minsky and Seymour Papert showed that perceptrons have limitations'—they cannot figure out non-linear problems, for example [22]. It eventually resulted in what's called the "AI winter," a time when people lost interest in neural networks.

The revival of neural networks started in the 1980s when it became possible to learn 'deep' (i.e., multi-layer) networks using the backpropagation algorithm (Rumelhart, Hinton, and Williams, 1986). Backpropagation became popular again thanks to improvements in computing hardware, reviving interest in deep learning. The concept of "deep learning" itself became popular in the 2000s when computer scientists began investigating models with many hidden layers, for example deep belief networks [16] and CNNs [19].

1.3.2 Mathematical Foundations

The point of deep learning is its theoretical foundations in linear algebra, calculus, probability, and optimization. Another potential application area of these mathematical methods is for the design, training, and testing of deep learning machines.

1. Linear Algebra: Deep learning uses operations with matrices, eigenvalue/eigenvector decomposition, and singular value decomposition. For example, neural networks take on a form where weights and biases are represented as matrices and vectors, respectively; in a feedforward pass, the multiplication of these matrices and application of activation functions occur [42].
2. Calculus: You'll need calculus if you want to tune the "knobs" of deep learning models. The optimization problem is solved by the backpropagation algorithm [25], which is applied to train neural networks, based on the chain rule of calculus for computing gradients of a loss function with respect to model parameters. These gradients are subsequently used to perform the parameter updates using optimization techniques such as stochastic gradient descent (SGD).
3. Probability and Statistics: Almost all deep learning models have a probabilistic interpretation used to model the uncertainty in predictions. For instance, Bayesian neural networks interpret model parameters as being drawn from probability distributions, enabling them to measure uncertainty [23]. In addition, maximum likelihood estimation and cross-entropy loss are inherent to deep model training.
4. Optimization: Optimization is the heart of deep learning—training a neural network means essentially minimizing some loss. Optimization methods including SGD and Adam [18], and RMSprop [27] are commonly used to train deep learning models.

1.3.3 Key Architectures in Deep Learning

There are myriad proposed deep learning architectures–all of which cater to different tasks and types of data. Listed below are several architectures:

1. Convolutional Neural Network (CNN): CNNs are designed to operate over grid-like data such as images. They use convolutional layers to extract spatial features and pooling layers for dimensionality reduction [20]. CNNs have been the driving force behind achieving state-of-the-art results in computer vision tasks, such as image classification, object detection and segmentation.
2. Recurrent Neural Networks (RNNs): It has been developed to deal with sequences data, time-series and natural language. RNN can model the temporal dependencies, and therefore can be used for tasks such as speech recognition and machine translation [17]. Nevertheless, RNNs are subject to vanishing gradients and long short-term memory (LSTM) networks [17] and gated recurrent unit (GRU) [12] were proposed as remedies.

3. Transformers: Transformers were introduced by [27], and revolutionized NLP as they replaced recurrent layers by self-attention. This architecture enables the parallel computation of sequences, and has been used for models such as BERT [13] or GPT [11], which obtained outstanding results in NLP tasks.
4. Generative Adversarial Networks (GANs): GANs were first proposed by [42], consisting of two competing neural networks: a generator and a discriminator. The generator creates synthetic data, while the discriminator attempts to distinguish real from fake. GANs have a wide range of applications such as image generation, style transfer, and data augmentation.

1.3.4 Applications of Deep Learning

Many successful deep learning applications have been developed in areas outside of business, and many believe these techniques will transform the way we conduct business. Some notable applications include:

1. Computer Vision: We are witnessing the dawn of computer vision with deep learning, e.g., face detection and recognition, self-driving cars, and medical X-ray analysis. For example, CNNs have demonstrated high accuracy in disease condition prediction using X-ray and MRI images [14].
2. Natural Language Processing: NLP changed a lot thanks to deep learning, with applications such as machine translation, sentiment analysis, and chatbots. In particular, transformers have achieved new state of the art in language modeling and text generation tasks [28].
3. Speech Recognition: Through deep learning, systems can be more precise in recognizing spoken words even under varying conditions such as accents, dialects, or background noise. In this field, RNNs and transformers are commonly used [15].
4. Reinforcement Learning: The fusion of deep learning and reinforcement learning is called deep reinforcement learning, where agents can learn difficult behaviors through interaction with the environment. This has been applied to achieve superhuman performance in games such as Go and chess [26].

1.4 Convolutional Neural Networks (CNNs)

Convolutional Neural Networks (CNN) are a specialized type of deep learning model that has sparked the successful story of computer vision. These networks are also particularly useful for problems such as image classification, object detection and segmentation. CNNs can automatically and adaptively learn spatial hierarchies of features from input images, and they are very effective for visual data processing. In this section, we present an overview of the CNNs with its structure, components and applications.

1.4.1 Architecture of CNNs

CNNs are inspired by the structure of the animal visual cortex, in which individual cells respond to overlapping regions of view. CNN is an ensemble of layers, and each layer conducts a specific manipulation on the input. The basiclayersin a CNN are the convolutional layers, pooling layers andfully connectedlayers. Figure shows the architecture of a Convolutional Neural Network (CNN) (Fig. 1.3).

- Convolutional Layers: The first layer of a CNN is a convolutional layer. It basically does the i/p image that is convolved with a learnable stes of filters (Or Kernels). Every filter convolves over the input image via taking the dot product with a local window of the input. The output is a feature map that indicates the presence of certain features in the input image. The convolution operation is defined as Eq. (1.1):

$$(f * g)(x, y) = \sum_{i=-\infty}^{\infty} \sum_{j=-\infty}^{\infty} f(i, j) \cdot g(x - i, y - j) \tag{1.1}$$

  where f and g are each the input image and filter and (x,y)(x,y) are spatial coordinates in the output feature map. After this convolution step an activation function is applied (e.g., Rectified Linear Unit or ReLU) to add nonlinearity to the model.
- Pooling Layers: Used to down-sampling the feature maps, reducing their running time and no. of parameters in the network. The most popular pooling operation is MaxPooling, which extracts the maximum value among local blocks of the feature map. Another kind of pooling is averagePool and globalAveragePooling. Pooling

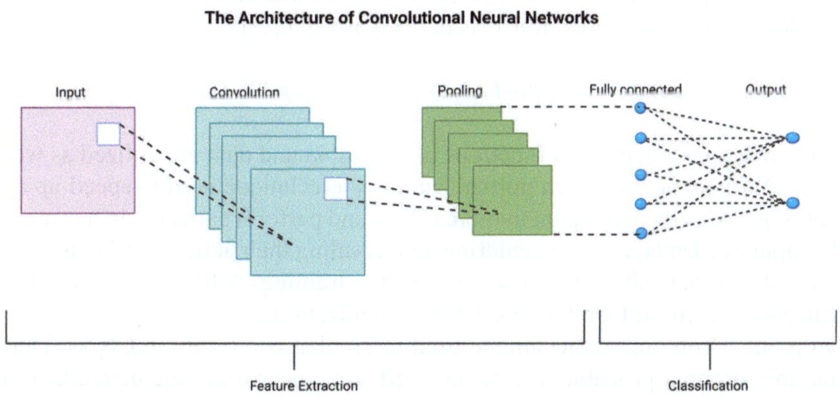

Fig. 1.3 The architecture of a Convolutional Neural Network (CNN). The process begins with the input image, followed by convolutional layers for feature extraction, pooling layers for dimensionality reduction, and fully connected layers that map the extracted features to the final classification output

layers also lead to invariance to small translations and, more generally, slightly distortions of the input image which contributes to a shift-invariant form.

- Fully Connected Layers: The high-level reasoning in the neural network takes place in the fully connected layers. These layers transform the flat feature maps from previous layers into the final output (e.g., class scores in a classification problem). Dense layers are typical to classical neural network, such that all the neurons in a layer are fully connected with all the neurons from its previous layer.

1.4.2 Key Components of CNNs

Many factors contribute to the success of CNN for visual recognition.

- Filter and Feature Maps: In CNNs, a filter is a small matrix that is convolved with an input image to detect specific features. Each filter have a purpose to patch out some particular feature like edge, texture or pattern. A feature map is a result of convolution operation which describes where the detected feature is in the image. For every convolutional layer, we employ multiple filters to learn various characteristics.
- Stride and Padding: Stride; the amount by which the filter shifts is called the stride. Stride is 1 means filter would move one pixel at a time and stride of 2 means it will jump 2% pixels. The spatial size of output feature map is determined by padding. Padding with zeros from both sides is included to ensure the size of image during filtering, as the filter calculates edges of an image.
- Activation Functions: Activators introduce non-linearity to the network and allow to learn complex patterns. The ReLU activation function is the most widespread in CNNs - easy and works. ReLU is defined as Eq. (1.2):

$$ReLU(x) = \max(0, x) \tag{1.2}$$

  Sometimes other activation functions like sigmoid and tanh are utilized as well.
- Batch Normalization: Batch normalization is a technique used to speed up deep learning CNNs by making them more robust and perform antically. It "normalizes the input of each layer by re-centering and rescaling their activations" to minimize internal covariate shift, which accelerates the training. A ReLU (as a non-linear activation function) then follows batch normalization.
- Drop out: Drop out is a technique used to combat overfitting in CNNs. During training, dropout probabilistically turns off some neurons in the network so that survivors will encode more robust features. This naturally enhances the generalization of the model.

1.4.3 Training CNNs

When training a CNN, one optimizes the parameters of the model to minimize a loss function that quantifies how different are the predicted and true outputs. The most widely adopted optimization approach in CNNs is (variants of) Stochastic Gradient Descent GPU architectures. The training procedure is as follows:

- Feed Forward: The input image is fed forward through the network and an output value is calculated with the current parameter settings.
- Loss Calculation: Loss function is calculated to quantify the difference between estimated output and true target.
- Backward propagation: Gradients of the loss function computed with respect to the model parameters are determined through calculus chain rule.

This training procedure is applied for multiple number of epochs to converge to a model with model parameters that minimize the value of the loss function.

1.5 Historical Development of Object Detection Techniques

Object detection has evolved from handcrafted features into deep learning methods that have become the standard for many applications. At the initial days of object detection, image-based approaches like Haar Cascade Classifier came into play that used simple and hand-crafted features (edges and corners) to localize objects. This was then succeeded by the sliding window approach, which transversely scans over an entire image with a fixed-size window at multiple position scales, and employs traditional classifiers such as Support Vector Machines (SVM) to find out potential objects. One important breakthrough was in 2005, when the Histogram of Oriented Gradients (HOG) descriptor together with linear SVM were employed for detecting people and other objects with much higher detection accuracy.

Deep learning, which emerged in the 2010s, is what really revolutionized object detection. Later on Convolutional Neural Networks (CNNs) were introduced to image classification and the potential for object detection, since CNNs could learn features automatically from data directly. In 2014, R-CNN (Region-based CNN) was proposed. It is a method to utilize CNNs on RoIs in an image that are determined by selective search.

This procedure was correct, but it was computationally superfluous. In 2015, Fast R-CNN was further developed to reduce the computation burden across regions and make it faster and more efficient. Faster R-CNN further removed selective search by adding a Region Proposal Network (RPN) which allowed end-to-end training of the network.

The next impact was made by single-shot detectors, such as YOLO (You Only Look Once) in 2016 and SSD (Single Shot MultiBox Detector) of the same year. Both of these methods formulated object detection as a single regression problem, where instead of predicting class labels with their location separately, the model output predicted bounding boxes and class labels directly from image input in one pass. This greatly increased the efficiency of object detection and it could be used for real applications in real time. More recently, RetinaNet (2017) proposed the Focal Loss for solving the issue of class imbalance, and is beneficial when detecting objects that are rare.

Later techniques would be a little bit more complex/advanced, e.g., Detection Transformers (DETR, 2020), where transformer-based architectures were used to model relationships between objects in the image. This method proposed the direct set prediction for object detection and achieved excellent detection performance and flexible receptive fields, but it needed to train on massive amount of data.

The difference between traditional object detection approaches and deep learning-based solutions is mostly in terms of feature extraction, speed, accuracy, training and scalability. Figure 1.4 shows the timeline of object detection model evolution (Table 1.1).

Feature-based object detectors suffered from the high “curse of dimensionality.” The range of features that provide a performance gain is determined by the knowledge or intuition of the domain expert. This may include texture, edge or corner features, gradient features, oriented gradients, and so on. In contrast, deep learning (DL) uses neural networks, in particular (convolutional neural networks (CNNs), that remove the requirement for feature definition, scale, and design. The network can define and learn features from data given. Hierarchical features arise from the layers of the network. Lower layer features may represent edges or gradients. Highest layer and deeper layer features may represent curves, shapes, and structures. Hierarchical features generalize better across object categories, variations, and domains. Empirical evaluation proves advantageous learning and emergence of these higher features over feature scrutability. Analysis of object detection signals does not provide an

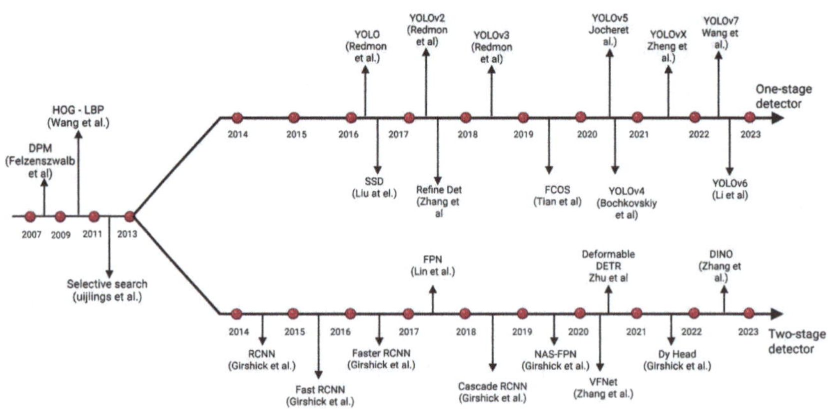

Fig. 1.4 Timeline of object detection model evolution. Major milestones in one-stage and two-stage detectors from 2007 to 2023, highlighting key frameworks such as HOG-LBP, DPM, RCNN family, SSD, YOLO variants, and more recent architectures

Table 1.1 Comparison of traditional versus deep learning-based object detection

Aspect	Traditional object detection	Deep learning-based object detection
Feature Extraction	Relies on handcrafted features like edges, corners, or textures (e.g., HOG, SIFT). Features are manually designed based on domain knowledge.	Uses CNNs to automatically learn features directly from image data. No need for manual feature extraction
Speed and Efficiency	Slower due to techniques like sliding windows, which require scanning the image multiple times at different scales.	Faster due to models like YOLO and SSD that perform detection in a single pass, enabling real-time applications
Accuracy	Limited by the quality of handcrafted features and often struggles with generalization across various object types and conditions.	High accuracy due to the ability of deep learning models to learn complex, hierarchical features from large datasets
End-to-End Training	Typically requires multiple steps such as feature extraction, region proposal, and classification, which are separate processes.	Models like Faster R-CNN and YOLO allow for end-to-end training, where all parts of the detection process are learned jointly
Scalability	Scaling traditional methods for new object types or environments often requires customization and manual tuning of features.	Deep learning models scale well with the availability of large datasets, improving performance with more data

appropriate "scrutability." DL approaches emerge to provide improved representations. Table 1.1 shows the comparison of traditional verses Deep Learning-Based Object Detection. Although learning based methods outperform hand designed, feature oriented representation, classical processing techniques such as sliding window scan the test image. Region proposal algorithms largely depend on scanning methods. Each window or region must be individually evaluated, leading to prohibitive inference times. These approaches demand, high Nd2 evaluations of the CNN. Current architectures require seconds to process a single image. When integrated into real-time application platforms such as autonomous vehicles, on-device execution environments, or live CCTV monitoring networks, an urgent, need for speed and efficiency has emerged. Current deep learning detection methods, including SSD, YOLO, and RetinaNet, perform single pass detections, where both regressing and classifying bounding box coordinates occur simultaneously over the entire image. These approaches substantially increase the speed and efficiency of detection methods over region proposal pipelines. Previous methods utilizing hand crafted techniques, including the use of Haar features and HOG plus cascade classifiers, do not generalize to unseen object classes or variations and struggle with high clutter and occlusions, large amounts of variation or complexity in object scale, position, orientation, or instantiation. The rise of deep network models has transformed a number of domains, including image classification, image segmentation, and speech recognition. With the advance of representation power of deep learning architectures across a broad range of image types, and the availability of large labeled datasets to facilitate supervised learning, deep network models in various forms have become the common denominator of state-of-the-art object detection methods. Prior to the emergence of deep learning, object detection systems involved the systematization of a sequence of components, usually including feature computation, data transformations, region proposal, classifiers, and spatio-temporal verification, that had to be optimized separately, ensuring each component operated well individually and dealt well with the output of the previous components. In the case of deep learning based object detection, such as the RPN network incorporated in Faster R-CNN, these components can be realized in a single computational model comprised of a neural network and its end-to-end training in a single optimization process. Another major effort to promote the use of deep learning for object detection was to eliminate the need for feature engineering to cope with new object classes and changing environments that had to be designed and optimized by human experts through experience over multiple iterations. Instead, networks can natively accommodate new classes and changing environments simply by training on large datasets from unintended environments using deep learning technologies. Faster R-CNN improves upon the earlier R-CNN ([1] and its successors) and relies on the use of a large single model similar to YOLO and SSD. Although YOLO and SSD models have comparable performances, Faster R-CNN also provided toy target results at the time of publication. The more heavily class-specific MIL based systems had considerably lower performance and greater customization to each novel task.

1.6 Importance in Computer Vision Applications

Object detection is a core task in computer vision because it serves as the foundational component for many practical implementations and use cases in various domains including: healthcare, security, transportation, retail and beyond. Object detection allows for the detection, classification and tracking of objects in images and videos enabling systems to intelligently interpret visual data (most times unseen) and automating processes that involve decisions [29].

One of the more prevalent use cases for object detection systems is the autonomous driving domain. Self-driving vehicles use object detection as an input to identify objects in the driver's domain, including but not limited to pedestrians, cyclists, road signs, lane markings, vehicle lights (e.g. red and green lights), and vehicles. Autonomous vehicles employ cameras and sensors along with object detection deep learning frameworks (e.g. YOLO (You Only Look Once) or Faster R-CNN) to manage large amounts of visual input, anticipate potential risks, and formulate driving responses [30, 31]. Object detection is one of the quintessential components of an autonomous driving system. In fact, without the ability to implement object detection, autonomous vehicle systems would not be able navigate consistently and effectively without compromising safety.

In medical imaging, object detection enhances the accuracy of diagnosis by detecting and classifying abnormalities in medical scans (for example, X-rays, MRIs and CT scans). AI-based object detection models support radiologists by detecting tumors, fractures, and other abnormalities within a scan with high accuracy. Deep learning-based models, such as RetinaNet and SSD (Single Shot MultiBox Detector), for automated cancer detection have been used for example, preventing missed diagnoses and increasing early detection rates [32, 33]. This application provides not only a quicker diagnosis, but also provides assurance that medical professionals can spend time focusing on more complex cases, thus improving healthcare delivery in general.

Object detection is relied upon in surveillance and security systems for real-time monitoring and threat detection. Smart surveillance cameras that are outfitted with object detection algorithms can detect suspicious activity, intrusion, or weapons from a large group of people. Additionally, these surveillance systems use facial recognition and person re-identification (Re-ID) technology to follow subjects across various camera feeds to support those in the law enforcement field with credible criminal investigations [34]. AI-based anomaly detection can alert security staff to abnormal behavior and reduce testing of unfounded risk to future security breaches.

Retailers apply object detection to enhance customer experiences and streamline business operations. Automated checkout systems in supermarkets leverage object detection to identify customer purchases and initiate the payment process without manually scanning items [35]. Likewise, object detection allows retailers to monitor inventory levels as stock items are moved on and off sale floors, thus increasing the efficiency of inventory management and reducing the possibility of overstock and stockout conditions. Object detection can also help retailers obtain behavioral tracking analytics about their customers, such as what products they purchase and how traffic patterns in the store are established. Retailers can leverage this type of data to allocate space more effectively and make decisions related to marketing strategies.

In manufacturing, object detection is critical for the quality control and automation processes. Computer vision systems using object detection algorithms can identify defective products along production lines, thus ensuring only high-quality products make it into the hands of consumers [36]. These systems also create efficiencies in the inspection process while reducing the time and distance an inspector has to travel and thus decreasing the risk of human error. Object detection is also useful for robotic automation manufacturing. For example, industrial robots rely on visual perception for both accurate assembly of products and efficient material handling.

The agricultural sector has also seen great advancements with object detection technology, particularly in precision farming. Drones and automated machinery that incorporate computer vision can monitor crop and soil conditions, detect pests, etc. [37]. Object detection can identify diseased plants quickly, so the farmer can act quickly and reduce crop losses. There are also systems that monitor livestock for movement detection via object detection-based technology; this will improve care for individual animals and help avoid disease spread in the herd.

Applications of augmented reality (AR) and virtual reality (VR) benefit tremendously via object detection because they improve interactivity. AR systems can recognize real-world objects and overlay digital information using object detection techniques [38]. For example, mobile applications that are AR-based can detect furniture pieces and let the user visualize how that piece would look in their home before purchase. VR gaming uses object detection to track hand gestures and motion that enhance the interactivity of the game to provide the user with the most immersive experience.

Object detection is making its mark in sports because of automated tracking systems that accurately track player movements, ball trajectories, and referee decisions [39]. Coaches gain incredible insights into player performance while referees are assisted with making accurate calls with video-assisted decisions and fans are engaged with live statistics and visualizations. Object detection is even making its way into sports broadcasting, where cameras can follow and track players automatically and provides for exciting, live coverage of events.

In robotics, object detection is often leveraged for intelligent automation. Service robots, such as delivery drones and robots used in warehouses, are now using object detection to navigate environments, avoid obstacles, and pick or place objects precisely [40]. Robotic-assisted surgery in hospitals use robotic systems with object detection that maximize precision in locating anatomical structures for surgeons, especially with complex surgical procedures. Humanoid robot systems that use object detection can also interact more naturally with humans by recognizing objects, gestures, and faces. Object detection is an essential component of smart city initiatives, in which intelligent infrastructure relies upon real-time data collection and analysis. For example, in traffic monitoring systems, object detection allows for vehicle identification, congestion measurement, and traffic flow optimization through real-time adaptive signal control [41]. By utilizing object detection, parking management systems can detect, identify, and allow reservation of available parking spaces, thereby lessening traffic caused by users searching for a parking space. Additionally, in public safety applications, object detection monitors crowd densities and allows for better crowd management practices during large events and evacuations.

In e-commerce, object detection enables a number of uses, ranging from product recommendation, visual search, and automated content moderation, to name a few. For instance, many online retailers directly incorporate object detection applications into their platforms to allow users to search for products with an image upload, rather than using traditional word-based search terms [42]. This helps users find the best product based on a visual search experience. E-commerce platforms also utilize object detection to moderate user-generated content, ensuring that inappropriate or counterfeit products do not appear in store listings. Object detection is a vital part of the entertainment industry, improving the capabilities of content creation and visual effects. Object detection is employed by directors and producers to produce appealing graphics, track an actor modelling a scene, and other editing tasks associated with specific object detection within a visual object [43]. Meanwhile, for video streaming companies, object detection relies on the highly visual nature of video to extend the definition of content and refine recommendations. Object detection is also an important component within the same visual frame in automatic captioning, and scene detection for improving accessibility to site functionality for disability [43].

Object detection is also frequently applied to environmental monitoring and disaster response. Remote imagery, with the assistance of AI analyses, has applied object detection to provide resource monitoring related to deforestation, animal populations, and other climate change impacts on the condition and state of the environment [44]. During natural disasters with ecological impacts like floods, earthquakes, and wildfires, object detection could also be applied to assess equity and capture search and rescue responses to find people safely. Object detection assists emergency response organizations in optimizing their respective functionalities to invoke quick and accurate conclusions when responding to disasters or emergencies.

Object detection, as part of computer vision, is being utilized in educational contexts as part of interactive learning environments and assistive technologies. For example, in smart classrooms, an AI system can detect objects, gestures, or facial expressions and adjust the teaching pedagogy according to levels of engagement [45]. Assistive technologies, for people who are blind or have low vision, use object detection as a way to describe their environment, read text present in images, and navigate using autonomy to promote accessibility and independence.

Object detection is also changing logistics and the supply chain in terms of tracking and automation. In warehouses, automated sorting systems now utilize object detection to identify and organize packages for efficient processing of logistics. Additionally, delivery drones or autonomous vehicles use object detection for identifying delivery locations which can enhance delivery services in what is known as the last mile portion of the supply chain. Finally, it is being utilized as part of supply chain monitoring as part of inspections of cargo, theft reduction, and safety inspections.

As a primary application within computer vision and working towards intelligent decision-making and automation, object detection is embedded and increasingly sophisticated in many different sectors. The ability to automatically detect, classify, and track objects in real-time is improving efficiency, safety, and user experience. Improvements in deep learning and AI technology have contributed to the evolution of object detection algorithms, and the increased use and sophistication of applications for object detection.

1.7 Object Tracking Versus Object Detection

Object detection and object tracking are two main tasks in computer vision with different goals and approaches. Although they are often used together, the two tasks are distinct in their aims, approaches, and uses. This paper provides a technical overview of both tasks and their distinctions and approaches under each task.

1.7.1 Object Detection: Techniques and Characteristics

Object detection involves detecting and localizing objects that appear in an image or video. Object detection encompasses two tasks, classification and localization. Classification is referring to classifying what type of object is in the image (i.e., person, car, dog) whereas localization refers to the location of that object in the image, usually by a bounding box.

1. Bounding Box Regression. Object detection algorithms typically use bounding box regression to predict the posed bounding box coordinates of an object. Intersection over Union (IoU) can then be computed to measure the predicted bounding box to the ground truth bounding box. IoU denotes the overlap between predicted and actual bounding boxes, and measures localization accuracy [46].
2. Sliding Window Approach. Classical object detection methods, such as Haar cascades, would implement a sliding window approach where the image would be scanned with a fixed sized window, and the classifier would be applied to each window to classify whether or not the window contained an object. This approach can be effective, however, it is computationally expensive for high resolution images [47].
3. Region Proposal Networks (RPNs): More contemporary object detection networks utilize Region Proposal Networks (RPNs) that then generate region proposals. The system works by predicting the location of objects by placing anchor boxes of diverse scales and aspect ratios on the image. RPNs help to reduce the computational burden with the efficiency of only examining a limited subset of regions where an object is likely to be instead of a full image [48].
4. Feature Extraction: The detection of an object relies significantly on feature extraction in separating an object from the back ground. A feature extractor, usually Convolutional Neural Networks (CNNs), is used to extract hierarchical features from the image. Features create a spatial hierarchy of features facilitating the detection of objects at a diversity of scales and orientations [49].
5. Non-Maximum Suppression (NMS): When a prediction is made, an additional post-processing step is then made by a process such as Non-Maximum Suppression (NMS) that suppresses bounding boxes that are overlapping spatially. NMS will choose the bounding box with the maximum confidence and suppress any other spatially overlapping boxes likely to be the same object. In this process only one bounding box to represent the object in the image at the end [50].

1.7.2 Object Tracking: Techniques and Characteristics

On the other hand, object tracking concerns tracking a specific object (or objects) across consecutive frames in a video sequence. The goal is to maintain the identity of the object over time as it moves, perhaps changes its appearance, or interacts with other objects.

1. Motion Estimation: Object tracking generally starts off with motion estimation, in which the tracking algorithm predicts the object's future position based on its previous position. Algorithms such as Kalman filters and particle filters are typically employed for motion estimation. Kalman filters provide a prediction on the object's state (position, velocity) by modeling the object with linear dynamic models, while particle filters use a set of particles to represent the object's potential state—making the filter more resilient to non-linear motion [51].
2. Appearance modeling: For tracking an object over time, the tracking algorithm has to model the object's appearance over time—not only the appearance might change due to lighting conditions, and occlusion of objects, but the object itself might deform. Many appearance modeling algorithms are based on representations, such as histograms, while other algorithms are based on deep learning (for example, using CNNs)—typically the first-class algorithms model appearance with various histograms based on pixel intensity, while the latter learning appearance which is invariant to its changes [52].
3. Data Association: In multi-object tracking, data association refers to the part of the algorithm that associates tracked objects over time between frames, often applying the Hungarian algorithm or Joint Probabilistic Data Association (JPDA) which specifically addresses the data association problem. The Hungarian algorithm assigns detections to trajectories while minimizing the overall cost associated with the assignment. JPDA is a more sophisticated method that takes into account the probability of each detection being associated with a trajectory in the presence of uncertainty [53].
4. Online tracking versus Offline tracking: Object tracking can be implemented online or offline. Online tracking is when each frame of a video is processed in consecutive order. This method is useful for real-time applications. Offline tracking processes a video sequence as a whole at once in search of the most accurate results. If the video is processed in the forward direction in time, the accuracy, especially of the initial detections, will be greater than if it were done online. Algorithms that are online, in particular those that utilize correlation filters, will typically have relatively low if not negligible computational complexity during the tracking step, but will not perform well in the presence of long occlusions. Offline algorithms especially, those that involve graph optimization methods may be able to better handle such difficult situations. However, there is an increase in computational complexity [54].
5. Occlusion Handling: Occlusion is one of the most challenging aspects of object tracking, and it occurs when the object temporarily disappears from view. Some occlusion handling techniques include re-identification and occlusion reasoning. Re-identification is attempting to match the object's appearance at time points before and after occlusion, whereas occlusion reasoning attempts to predict an object's location while occluded using motion and context [55].

1.7.3 Key Differences Between Object Tracking and Object Detection

1. Temporal Consistency. Object tracking is unique in that it is focused on maintaining temporal consistency in terms of tracking an object across frames, while object detection is, by definition, frame independent—it is about simply labeling an object in a frame and does not inherently consider the object's motion across frames. Although this distinction may seem trivial, it can be critical in applications such as surveillance, where following an object's motion across frames is more desirable than simply detecting it in a particular frame [56].
2. Computational Complexity. Object detection is often considerably more computationally-intensive than object tracking, primarily in the utilization of deep learning methods. An object detection algorithm considers the entire image to find the location of all accessible objects, while a tracking algorithm only considers where it is predicted that net object will be located in the portion of the image derived from the location from the previous frame. In light of this, tracking is much faster, making it more desirable for real-time applications [57].
3. Occlusions and Appearance Changes. Occlusions, or changes in object appearance, are considered when developing object tracking algorithms whereas this is less of a concern when developing object detection algorithms. An object detector is purposely designed under the condition the object is completely visible in the frame, while an object tracking algorithm assumes the object is present and can account for partially occluded objects, alterations in light, or other changes to its appearance over a duration of time [58].
4. Data Association: Data association is a vital part of object tracking, which we need to think about because we need to link objects frame-by-frame which allows the algorithm to keep track of the same object as it moves through the scene. In contrast with detection, this is not something that we have to worry about because the detectors will analyze each independent frame independently. Unfortunately, the idea of introducing data association can be even more problematic with multiple objects potentially crossing each other's paths or occluding one another [59].
5. Application Scenarios: Object detection is usually used for tasks, in which the intention is to detect and localize objects in a single frame (e.g., image classification, self-driving cars and facial recognition). Object tracking is typically used for applications that need to follow objects through time (e.g., video surveillance, sports analysis, human-computer interaction) [60].

Object detection and object tracking are distinct but complementary tasks in the area of computer vision. Object detection aims to identify and localize objects in images for individual frames, using techniques such as bounding box regression, region proposal networks and non-maximum suppression. Object tracking focuses on preserving the identity of an object through individual frames of a video using techniques such as motion estimation, appearance models, and data association. Although the two tasks may share some techniques (e.g., feature extraction using CNNs).

1.8 Book Outline

In this chapter, the book commences with an introduction, providing a foundational overview of object detection and localization. This chapter explores the historical development of object detection techniques, the importance of deep learning in computer vision, and the distinction between object tracking and object detection. By establishing key concepts, it sets the stage for the in-depth discussions and advancements presented in subsequent chapters.

Chapter 2, Image Processing and Preprocessing, delves into the essential steps required to prepare data for object detection tasks. This chapter explores image representation formats, data annotation techniques, and preprocessing strategies. Additionally, it examines synthetic data generation and methods for handling imbalanced datasets, ensuring a solid foundation for training deep learning models.

Chapter 3 introduces a comprehensive exploration of object detection algorithms, encompassing both single-stage and two-stage detectors. It discusses the evolution of models such as YOLO, SSD, and EfficientDet, alongside advanced architectures like Faster R-CNN, Mask R-CNN, and Feature Pyramid Networks. This chapter provides a deep dive into how these models enhance accuracy and efficiency in object detection applications.

Chapter 4, Understanding Localization, investigates the methodologies used to precisely identify object positions within an image. It covers bounding box regression, keypoint detection, and segmentation-based approaches. This chapter also introduces essential evaluation metrics such as Intersection over Union (IoU) and Mean Average Precision (mAP), offering a structured approach to assessing model performance.

Chapter 5 explores real-world applications of object detection and localization across diverse domains, including autonomous driving, surveillance systems, medical imaging, and robotics. By examining case studies and implementation strategies, this chapter highlights the transformative impact of deep learning in these fields and underscores its practical significance.

Chapter 6 delves into the integration of deep learning frameworks and libraries used for object detection. It presents an overview of TensorFlow, PyTorch, MATLAB, OpenCV, and MXNet, providing insights into their advantages and applications. This chapter also explores the use of pre-trained models, model compression techniques, and optimization strategies for efficient deployment.

Chapter 7 propounds innovative techniques for training and fine-tuning object detection models. It explores transfer learning, hyperparameter optimization, and loss functions while addressing critical issues such as overfitting and underfitting. By focusing on best practices, this chapter provides a structured approach to enhancing model performance.

Chapter 8 investigates the current challenges in object detection, from handling occlusions and small object detection to real-time processing constraints. It also delves into ethical considerations related to AI deployment, including bias, privacy concerns, and the responsible use of deep learning in critical applications.

Chapter 9 concludes the book, summarizing key findings and insights from previous chapters. It provides a forward-looking perspective on the future of object detection and localization, discussing emerging trends and potential advancements in deep learning methodologies. Recommendations for future research and development initiatives are also presented, offering guidance for continued innovation in the field.

References

1. Girshick, R., Donahue, J., Darrell, T., Malik, J.: Rich feature hierarchies for accurate object detection and semantic segmentation. In: Proc. IEEE Conf. Comput. Vis. Pattern Recognit. (CVPR) (2014)
2. Ren, S., He, K., Girshick, R., Sun, J.: Faster R-CNN: Towards real-time object detection with region proposal networks. In: Adv. Neural Inf. Process. Syst. (NeurIPS) (2015)
3. Redmon, J., Divvala, S., Girshick, R., Farhadi, A.: You only look once: Unified, real-time object detection. In: Proc. IEEE Conf. Comput. Vis. Pattern Recognit. (CVPR) (2016)
4. Liu, W., Anguelov, D., Erhan, D., Szegedy, C., Reed, S., Fu, C.Y., Berg, A.C.: SSD: Single shot multibox detector. In: Eur. Conf. Comput. Vis. (ECCV) (2016)
5. Lin, T.Y., Dollár, P., Girshick, R., He, K., Hariharan, B., Belongie, S.: Feature pyramid networks for object detection. In: Proc. IEEE Conf. Comput. Vis. Pattern Recognit. (CVPR) (2017)
6. Carion, N., Massa, F., Synnaeve, G., Usunier, N., Kirillov, A., Zagoruyko, S.: End-to-end object detection with transformers. In: Eur. Conf. Comput. Vis. (ECCV) (2020)
7. He, K., Gkioxari, G., Dollár, P., Girshick, R.: Mask R-CNN. In: Proc. IEEE Int. Conf. Comput. Vis. (ICCV) (2017)
8. Zhou, X., Wang, D., Krähenbühl, P.: Objects as points. arXiv preprint arXiv:1904.07850 (2019)
9. Tan, M., Le, Q.V.: EfficientNet: Rethinking model scaling for convolutional neural networks. In: Int. Conf. Mach. Learn. (ICML) (2019)
10. Long, J., Shelhamer, E., Darrell, T.: Fully convolutional networks for semantic segmentation. In: Proc. IEEE Conf. Comput. Vis. Pattern Recognit. (CVPR) (2015)
11. Brown, T.B., et al.: Language models are few-shot learners. In: Adv. Neural Inf. Process. Syst. 33, 1877–1901 (2020)
12. Cho, K., et al.: Learning phrase representations using RNN encoder-decoder for statistical machine translation. arXiv preprint arXiv:1406.1078 (2014)
13. Devlin, J., et al.: BERT: Pre-training of deep bidirectional transformers for language understanding. arXiv preprint arXiv:1810.04805 (2019)
14. Esteva, A., et al.: Dermatologist-level classification of skin cancer with deep neural networks. Nature **542**(7639), 115–118 (2017)
15. Graves, A., Mohamed, A., Hinton, G.: Speech recognition with deep recurrent neural networks. In: IEEE Int. Conf. Acoust. Speech Signal Process. (ICASSP), pp. 6645–6649 (2013)
16. Hinton, G., Osindero, S., Teh, Y.W.: A fast learning algorithm for deep belief nets. Neural Comput. **18**(7), 1527–1554 (2006)
17. Hochreiter, S., Schmidhuber, J.: Long short-term memory. Neural Comput. **9**(8), 1735–1780 (1997)
18. Kingma, D.P., Ba, J.: Adam: A method for stochastic optimization. arXiv preprint arXiv:1412.6980 (2015)
19. LeCun, Y., Bengio, Y., Hinton, G.: Deep learning. Nature **521**(7553), 436–444 (2015)
20. LeCun, Y., et al.: Gradient-based learning applied to document recognition. Proc. IEEE **86**(11), 2278–2324 (1998)
21. McCulloch, W.S., Pitts, W.: A logical calculus of the ideas immanent in nervous activity. Bull. Math. Biophys. **5**(4), 115–133 (1943)

22. Minsky, M., Papert, S.: Perceptrons: An Introduction to Computational Geometry. MIT Press, Cambridge (1969)
23. Neal, R.M.: Bayesian learning for neural networks. PhD thesis, University of Toronto (1995)
24. Rosenblatt, F.: The perceptron: A probabilistic model for information storage and organization in the brain. Psychol. Rev. **65**(6), 386–408 (1958)
25. Rumelhart, D.E., Hinton, G.E., Williams, R.J.: Learning representations by back-propagating errors. Nature **323**(6088), 533–536 (1986)
26. Silver, D., et al.: Mastering the game of Go without human knowledge. Nature **550**(7676), 354–359 (2017)
27. Tieleman, T., Hinton, G.: Lecture 6.5-rmsprop: Divide the gradient by a running average of its recent magnitude. COURSERA: Neural Networks for Machine Learning (2012)
28. Vaswani, A., et al.: Attention is all you need. In: Adv. Neural Inf. Process. Syst. 30, 5998–6008 (2017)
29. Redmon, J., Divvala, S., Girshick, R., Farhadi, A.: You only look once: Unified, real-time object detection. In: Proc. IEEE Conf. Comput. Vis. Pattern Recognit. (CVPR), pp. 779–788 (2016)
30. Ren, S., He, K., Girshick, R., Sun, J.: Faster R-CNN: Towards real-time object detection with region proposal networks. In: Adv. Neural Inf. Process. Syst. 28, 91–99 (2015)
31. Lin, T.Y., Goyal, P., Girshick, R., He, K., Dollár, P.: Focal loss for dense object detection. In: Proc. IEEE Int. Conf. Comput. Vis. (ICCV), pp. 2980–2988 (2017)
32. Liu, W., Anguelov, D., Erhan, D., Szegedy, C., Reed, S., Fu, C.Y., Berg, A.C.: SSD: Single shot multibox detector. In: Eur. Conf. Comput. Vis. (ECCV), pp. 21–37 (2016)
33. Zhao, Z.Q., Zheng, P., Xu, S.T., Wu, X.: Object detection with deep learning: A review. IEEE Trans. Neural Netw. Learn. Syst. **30**(11), 3212–3232 (2019)
34. Zhang, Y., Wang, Z., Liu, H.: Object detection in retail: A survey. IEEE Trans. Pattern Anal. Mach. Intell. **42**(10), 2130–2145 (2020)
35. Wang, L., Zhang, J., Wang, Y.: Object detection in manufacturing: A review. Robot. Comput. Integr. Manuf. **68**, 102–112 (2021)
36. Kamilaris, A., Prenafeta-Boldú, F.X.: Deep learning in agriculture: A survey. Comput. Electron. Agric. **147**, 70–90 (2018)
37. Azuma, R., Baillot, Y., Behringer, R., Feiner, S., Julier, S., MacIntyre, B.: Recent advances in augmented reality. IEEE Comput. Graph. Appl. **21**(6), 34–47 (2011)
38. Cioppa, A., Deliege, A., Van Droogenbroeck, M.: A bottom-up approach based on semantics for action recognition in soccer videos. IEEE Trans. Circuits Syst. Video Technol. **30**(9), 3156–3170 (2020)
39. Siciliano, B., Khatib, O.: Springer Handbook of Robotics. Springer, Heidelberg (2016)
40. Alam, F., Mehmood, R., Katib, I., Albeshri, A.: Smart cities: A survey on data management, security, and enabling technologies. IEEE Commun. Surv. Tutor. **22**(1), 271–332 (2020)
41. He, K., Zhang, X., Ren, S., Sun, J.: Deep residual learning for image recognition. In: Proc. IEEE Conf. Comput. Vis. Pattern Recognit. (CVPR), pp. 770–778 (2016)
42. Goodfellow, I., Bengio, Y., Courville, A.: Deep Learning. MIT Press, Cambridge (2016)
43. Li, X., Wang, W., Hu, X., Yang, J.: Selective kernel networks. In: Proc. IEEE/CVF Conf. Comput. Vis. Pattern Recognit. (CVPR), pp. 510–519 (2020)
44. Chen, X., Xie, H., Hwang, G.J.: AI-powered object detection for interactive learning in smart classrooms. Comput. Educ. **168**, 104–117 (2021)
45. Bogue, R.: Growth in e-commerce boosts innovation in the warehouse robot market. Ind. Robot **45**(6), 705–708 (2018)
46. Everingham, M., Van Gool, L., Williams, C.K.I., Winn, J., Zisserman, A.: The Pascal Visual Object Classes (VOC) Challenge. Int. J. Comput. Vis. **88**(2), 303–338 (2010)
47. Viola, P., Jones, M.: Rapid object detection using a boosted cascade of simple features. In: Proc. IEEE Conf. Comput. Vis. Pattern Recognit. (CVPR), vol. 1, pp. I–I (2001)
48. Ren, S., He, K., Girshick, R., Sun, J.: Faster R-CNN: Towards real-time object detection with region proposal networks. In: Adv. Neural Inf. Process. Syst. 28, 91–99 (2015)
49. LeCun, Y., Bengio, Y., Hinton, G.: Deep learning. Nature **521**(7553), 436–444 (2015)

50. Neubeck, A., Van Gool, L.: Efficient non-maximum suppression. In: Int. Conf. Pattern Recognit. (ICPR), vol. 3, pp. 850–855 (2006)
51. Kalman, R.E.: A new approach to linear filtering and prediction problems. J. Basic Eng. **82**(1), 35–45 (1960)
52. Comaniciu, D., Ramesh, V., Meer, P.: Kernel-based object tracking. IEEE Trans. Pattern Anal. Mach. Intell. **25**(5), 564–577 (2003)
53. Kuhn, H.W.: The Hungarian method for the assignment problem. Nav. Res. Logist. Q. **2**(1–2), 83–97 (1955)
54. Zhang, L., Li, Y., Nevatia, R.: Global data association for multi-object tracking using network flows. In: Proc. IEEE Conf. Comput. Vis. Pattern Recognit. (CVPR), pp. 1–8 (2008)
55. Yang, B., Nevatia, R.: An online learned CRF model for multi-target tracking. In: Proc. IEEE Conf. Comput. Vis. Pattern Recognit. (CVPR), pp. 2034–2041 (2012)
56. Yilmaz, A., Javed, O., Shah, M.: Object tracking: A survey. ACM Comput. Surv. **38**(4), 13–es (2006)
57. Smeulders, A.W.M., Chu, D.M., Cucchiara, R., Calderara, S., Dehghan, A., Shah, M.: Visual tracking: An experimental survey. IEEE Trans. Pattern Anal. Mach. Intell. **36**(7), 1442–1468 (2014)
58. Li, X., Hu, W., Shen, C., Zhang, Z., Dick, A., van den Hengel, A.: A survey of appearance models in visual object tracking. ACM Trans. Intell. Syst. Technol. **4**(4), 1–48 (2013)
59. Milan, A., Leal-Taixé, L., Reid, I., Roth, S., Schindler, K.: MOT16: A benchmark for multi-object tracking. arXiv preprint arXiv:1603.00831 (2016)
60. Wu, Y., Lim, J., Yang, M.H.: Online object tracking: A benchmark. In: Proc. IEEE Conf. Comput. Vis. Pattern Recognit. (CVPR), pp. 2411–2418 (2013)

Chapter 2
Image Processing

Abstract This chapter discusses the important image processing and preprocessing stages that are essential stages in developing a strong object detection system. It starts with a discussion of digital image representation, looking at binary images, grayscale and color images in the RGB and HSV models, and then discusses common file formats (JPEG, PNG, TIFF) including their impact on processing efficiency and accuracy in models. The chapter will then discuss data annotation, which plays an important role, including techniques using bounding boxes and polygons, to keypoint annotation and instance segmentation, and discusses tools including LabelImg and CVAT that can be used to create high-quality labeled datasets. A major concern of this chapter shall be data augmentation techniques, which will be broadly organized into geometric transformations, color space transformations, and advanced techniques, such as CutMix and GANs that may be used to artificially expand a training dataset, thus improving generalization and mitigating overfitting. Finally, this chapter will wrap up with important preprocessing steps such as image normalization, resizing, denoising, and balancing annotations which are important for efficiently preparing a dataset for consumption by a model. This chapter discusses the challenges associated with unbalanced datasets and sets forth methods such as SMOTE and focal loss, as well as the exciting opportunity of synthetic data generation as a new and effective method of dealing with unbalanced datasets, as well as issues of scarcity and privacy. In general, the chapter provides practitioners with the methods to develop effective data pipelines, resulting in effective input data that is trained to be reliable and accurate in object detection.

2.1 Image Representation and Formats

Digital images are the rudimentary foundation of deep literacy-based object discovery and localization systems. There is a passion for understanding how images represent things and how they are stored, because the colorful formats influence both data speed and model quality. A digital image is a grid of pixels, and each pixel has numbers that determine its color and brightness. The simplest form of a double image

A. Elhanashi and S. Saponara, *Deep Learning for Object Detection and Localization*,
https://doi.org/10.1007/978-981-95-4846-0_2

uses two values: 0 (black) and 1 (white). Double focusing images are effective for shape finding but not provide sufficient detail for recognizing developed objects [3]. Grayscale images use only one intensity channel, which goes from 0 (black) to 255 (white). Such images reduce the computation required but eliminate color information that is vital for certain kinds of discovery [9]. As object discovery tasks mature, color images are preferred since they hold more information. The most common digital color model is RGB, in which each pixel is defined by three intensities of the primary colors red, green, and blue. HSV (Hue, Saturation, Value) color model splices brightness factors away from color, making it more robust to variations in lighting [8]. Most videotape processing uses the YUV/YCbCr color space, which separates brightness (luminance) from color (chrominance) so that the data can be compressed efficiently [7].

Image file format however also holds great significance in deep learning pipelines. JPEG (Joint Photographic Experts Group) happens to be one of the most commonly used formats because it quite efficiently employs lossy compression that reduces file size significantly. Though, this compression can introduce artifacts that could be detrimental to accuracy in modeling [10]. While PNG (Portable Network Graphics) uses lossless compression, image quality is not sacrificed and no artifacts are introduced. PNG files support transparency which adds value for some detection and segmentation applications [1]. For certain high-quality applications such as medical imaging or satellite data analysis, TIFF (Tagged Image File Format) is typically used since the storage is lossless; however, this can hardly be applied in real time because of the huge file sizes created [6].

New types such as WebP give a good mix between how well they compress and quality of image, usually better than JPEG and PNG when it comes to making files smaller but keeping the look the same. This has caused more use in deep learning data sets [4]. At the same time, RAW types, which keep unchanging sensor info from top cameras give most detail but need change before being used in deep learning models [2]. Image representation and format play a crucial role in computational efficiency and model accuracy. Using compressed formats like JPEG will consume less memory and process faster but will lose fine details needed for accurate object localization.

Though lossless formats, PNG and TIFF keep the edges and surface of an object quite clear, which supports improvement in accuracy but increases storage as well as processing [5]. This is also true for RAW files because before they are given to convolutional neural networks (CNNs), the raw files go through demosaicing during preprocessing unlike normal RGB images that are typically ready for application. This makes image representation and format very critical for deep learning in object detection. JPEG serves the trade-off between size and speed while PNG gets better quality in cases where precision edge detection is important. The next big thing may be WebP and high dynamic range (HDR) formats taking a chunk of the improvements with respect to efficiency and quality trade-offs. Understanding all these basics should enable researchers and practitioners to create better-preprocessing pipelines as well as improve model performance. Figure illustrates the process of converting a grayscale image into its corresponding data matrix and 2D tensor representation (Fig. 2.1).

Fig. 2.1 Visualization of the process of converting a grayscale image sample into its corresponding data matrix and 2D tensor representation. The left panel displays a simplified grayscale image, which is then transformed into normalized pixel values forming the image data matrix (middle panel). The right panel shows the resulting image tensor, including details about its rank, shape, and values

2.2 Data Annotation

It's in data annotation that the very base and indeed all the most critical steps will be found in training deep learning models for object detection and localization since such annotation puts a premium on the model's ability to learn recognizing or localizing objects within an image or video frame. Classification alone is thus, using a label that denotes simply the presence of an object, insufficient even for arbitrary object-detection tasks, which require spatial information—bounding boxes or segmentation masks to define an object's exact position and its boundaries. The quality, consistency, and granularity of those annotations will define the generalization capability of a model tọ the real world; hence, such an integral phase of the machine learning pipeline renders data annotation indispensable especially in the light of huge high volume annotated ground-truth data sets for training supervised learning models. For such learning models, an annotated data set serves as the ground truth on which they rely during their training work. In object detection, even the smallest annotation errors, such as bounding boxes misaligned with objects, incorrectly assigned class labels, or even totally missing objects in the annotation, propagate through the learning process and lead to poor generalization, increased false positives, or localization inaccuracy in deployed systems. Case in point-the mislabeling of a pedestrian or miscoding of a traffic sign in autonomous driving could spell disaster, which only goes on to underline the very critical importance of precision and consistency in the annotations. Figure 2.2 illustrates the overview of the five primary types of data annotations in deep learning, which includes: image annotation, audio annotation, text annotation, geospatial annotation, and video annotation.

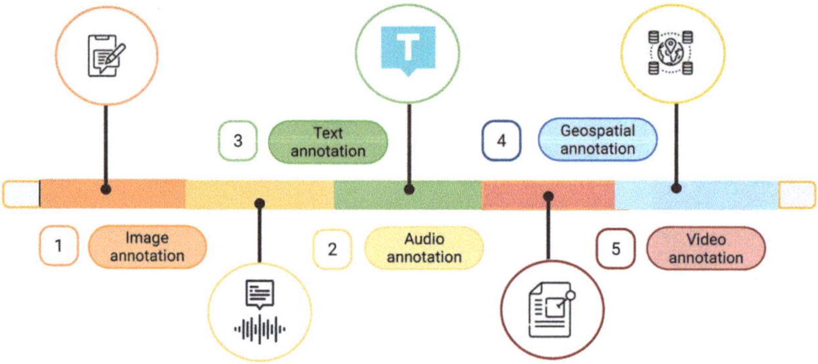

Fig. 2.2 Overview of the five primary types of data annotation in deep learning: image annotation, audio annotation, text annotation, geospatial annotation, and video annotation. Each annotation type is represented with a distinct color and icon, illustrating their respective positions within the overall data annotation workflow

2.2.1 Types of Data Annotation for Object Detection

For effective model training, proper annotations for object detection and localization are necessary. The following are the most commonly used types of an annotation:

- Bboxes: Bounding boxes are rectangular annotations that surround objects in an image. They provide a format representation for coordinates typically as xmin, ymin, xmax, ymax to set up space location for an image. They are most apparent, clear, and computable format for object detection. Examples of benchmark datasets include COCO [18]; Pascal VOC [12]. Bounding boxes are speedily annotated but not perfectly successful with different objects such as trees, furniture, occluded items, reduced numbers of accurate localization caused by background pixels, etc. So few frameworks have developed rotated bounding boxes (e.g., text detection) or objectness score refinement [19]. In spite of the above shortcomings, it is to the base to train detectors like YOLO and Faster R-CNN, but mainly in real-time applications.
- Polygon annotation outlines the exact contours of an entity and provides pixel-perfect boundaries for much-improved localization as opposed to rectangular bounding boxes. The best form of annotation for objects that are not easily contained in a bounding box format (for example, trees or animals or a tumor in medical imaging) is polygon annotation as it gives much more exact annotation without all that background noise or gaming. This allows the model to learn fine-grained spatial features while reducing the number of false positives in clutter because of space like other scene clutter. Because of this reason, polygon annotation is used in automated driving (precise lane detection), tumor segmentation in medical imaging, aerial surveillance or building footprint mapping. Basically, it is a very time-consuming process in comparison to bounding boxes using specialized tools (CVAT, LabelMe) and skilled users for consistency in annotation.

- Semantic segmentation is a pixel-wise annotation method in which a class label or category (like "road", "car", or "pedestrian") is assigned to each and every pixel of the image for an in-depth comprehension of the scene [21]. In contrast to bounding boxes that deliver only coarse localization, semantic segmentation is capable of resolving object boundaries very precisely, thus becoming essential and indispensable for applications that require a high degree of spatial accuracy; these include autonomous driving, medical imaging, and land cover mapping. Typically modern techniques apply fully convolutional networks (FCNs) and U-Net architectures for the purposes of real-time segmentation. The drawbacks mainly stem from the computational demand involved and the requirement for extensive labeling of datasets in order to avoid class imbalance and edge case errors.
- In Keypoint Annotation marks unique points in and around an object (for example, facial landmarks, joints in human pose estimation, or part locations in animal morphology). This annotation method is of significant importance for tasks that require fine-grained localization, as relationships between points would matter in the structure. In case of example, facial recognition, keypoints are the eyes, nose, and mouth to make a judgment assessment about facial expressions or alignment. For sports analytics, joint key points monitor the movements of athletes associated with performance assessment. Unlike bounding boxes, keypoints allow for spatial information to be mapped with a higher precision, hence are likely considered for applications in medicine (e.g., identifying bone landmarks) and robotics (e.g., object manipulation). Annotation as such becomes very problematic with occlusions or deformations by the object and therefore requires validation techniques which are robust.
- Instance Segmentation: This is the most advanced annotation technique which involves a combination of object detection and semantic segmentation in that it not only classifies every pixel but also distinguishes between each individual instance of an object [14]. Instance segmentation does everything semantic segmentation does, but for each object it includes a distinguishing name, for example Car A, Car B, etc. This shows how instance segmentation will feature in applications requiring highly accurate object interaction analysis like in autonomous driving (pedestrian tracking), medical imaging (tumor delineation), and robotics (object manipulation). Popular models like Mask R-CNN use instance segmentation to achieve the highest levels of accuracy in localizing while efficiently using computational resources.

2.2.2 Annotation Tools and Platforms

The annotation of necessary data is very important for a deep learning model to learn accurately, particularly in the area of object detection and localization. Thoroughly annotating the data requires strict control on precise labeling techniques, standardized protocols, and job of annotation due to minimize errors that might enter. As the dataset size and complexity increase, it becomes more important to have scalable annotation workflows for maintaining the model performance evaluation in a consistent and reliable way. Some tools allow efficient data annotation for deep learning:

- LabelImg (Open-source)—Lightweight graphical tool used for the fast annotation of bounding boxes in object detection tasks. Numerous formats are supported (Pascal VOC, YOLO, and COCO JSON), with the possibility of fast labeling using shortcut keys, which is especially handy for small-to-medium projects [15]. It has risen as a choice of preference among the researchers and hobbyists due to its simplicity and cross-platform compatibility (Windows/Linux/macOS).
- CVAT (Computer Vision Annotation Tool) was created by Intel and is an open-source, web-based platform that supports multiple annotation types such as bounding boxes, polygons, polylines, and semantic segmentation. It is a collaborative tool that allows for annotating, video frame interpolation, and AI-assisted auto-annotation, which is a perfect fit for large-scale computer vision projects [16]. It also provides a REST API to integrate easily into any machine learning pipeline.
- VGG Image Annotator (VIA): It is a web-based lightweight annotation tool developed at the Visual Geometry Group, Oxford which supports polygonily segmenting, marking keypoints, and bounding box annotation [11]. It is completely client site-based and does not require any installation or dependencies on any server, unlike many of its commercial counterparts. This makes it highly suitable for collaborative projects with privacy constraints because it exports annotations using standard formats (JSON, CSV), compatible with the major deep learning frameworks like TensorFlow and PyTorch. VIA, being an open-source and customizable tool, has gained popularity with researchers concerned about fine-grained object localization, especially in medical imaging and wildlife conservation [11]. Resource consumption is also very low, which means it can be easily used on devices with low capacity, thereby filling a major gap in ensuring accessibility of annotation.

2.2.3 Challenges in Data Annotation

Despite advancements, data annotation presents several challenges which include:

- Scalability: The manual annotation of vast data amounts takes such resources: time, money, and humans, as it takes hundreds of hours of time in human efforts. It's impossible to label manual datatypes growing from hundreds of thousands to millions of images (for example, for autonomous vehicles and satellite imagery). Human labeling then becomes unreasonable. Some solutions like AI-assisted annotation tools, crowdsourcing, and, say, semi-supervised learning, help control costs while ensuring quality annotations.
- Annotation Consistency: Variability due to subjective interpretations from different annotators, ambiguous labelling guidelines, or human errors may introduce

label noise that can degrade model performance. Consequently, mechanisms to diminish such label noise through IAA metrics, standardized protocols, and iterative review cycles are essential to certifying the reliability of training data, especially for such mission-critical applications as medical imaging or autonomous driving.

- Edge Cases: Obscuring or obstructing the view of an object increases the difficulty of annotating it correctly. These situations encumber human annotators as much as they do the automated labeling systems since the actual edges and sometimes even existence of an object become ambiguous. For example, ambiguous annotations may arise for a pedestrian occluded partly behind a vehicle or a traffic sign with part of it covered up. Guidelines for handling such edge cases must be insisted upon clearly (e.g., "label only visible areas") and iteratively reviewed to preserve the dataset's integrity. An emphasis on model validation and data augmentation would help boost generalization, particularly since models trained with imperfect annotations might poorly perform in the real world.

2.2.4 *Best Practices for High-Quality Annotations*

To ensure reliable annotations, researchers and practitioners should follow these best practices:

- Clearly defined Annotation Guidelines are essential to maintaining the consistency and quality of labeling datasets, especially with such arduous cases as those concerning truncated, occluded, or overlapping objects. This means the guidelines should define sharp rules when it comes to dealing with edge cases: for example, whether partially visible objects should be labeled at all; what should be regarded as adjacent instances of the same class; what are the minimum size thresholds for small or blurry objects; and class-specific distinctions (e.g., differentiating "person" from "rider" in autonomous driving datasets). Well-articulated guidelines serve to minimize inter-annotator variability by standardizing such decisions, thus minimizing labeling flaws and increasing the overall reliability of the dataset for robust model training.
- Inter-Annotator Agreement (IAA) is a measure of the consistency of labeling among annotators by various statistical metrics like Cohen's Kappa or Fleiss' Kappa. The higher the agreement, the clearer the guidelines were; the lower the score, the more ambiguous the instructions were and the higher the need for them to be corrected. IAA is important to ensure the quality of benchmark datasets, helping to minimize annotation noise that could otherwise induce bias into a machine learning model. IIAs should regularly be checked to ensure a better trustworthiness of the annotation process when working on subjective or complex labeling assignments.
- The Iterative Review Process is vital in data annotation quality assurance, employing rounds of verification to detect and correct labeling mistakes. This cycli-

cal annotation-review-refinement process is engaged in first making annotations, then via feedbacks from reviewers refining those annotations in successive rounds toward consistency and accuracy. As discrepancies are systematically reconciled, the process effectively reduces noise in the training data, improving model reliability and performance on object detection tasks.
- Active Learning: is a strategic approach in which certain samples are selected as candidates for relabeling and not all samples are indiscriminately relabeled. Instead one does this by learning from the most important examples – the ones for which model predictions are least confident. This method reduces labeling costs and hugely increases the performance of models [20]. Active learning is very useful now a days in several large scale datasets, which are cost-intensive for manual annotation for building better models but with lesser labeled samples. The same is used widely in medical imaging, autonomous driving, and NLP, where the cost incurred for obtaining precise annotations is very high.

2.3 Data Augmentation Techniques

Data augmentation is the most underrated part of deep learning and computer vision, especially when working with limited datasets. It artificially expands the training dataset with various transformations done on the original images for better generalization by the model and reduced overfitting [24]. Broad categories of data augmentation techniques include geometric transformation, color space augmentation, kernel filtering, mixing images, and adversarial techniques [17]. The following is their detailed explanation with the benefits attached to each application. Figure 2.3 shows the illustration of common geometric transformation of data augmentation.

2.3.1 Geometric Transformations

Geometric transformations modify any characteristic of a geometric figure such as position, size, or shape. Among the many varieties of geometric transformations are translation (the act of shifting), rotation (the act of turning), scaling (the act of resizing), and reflection (the act of flipping). These transformations are at the heart of computer graphics, image processing, and engineering, helping to manipulate and analyze objects that perform in either 2D or 3D space. Among the common set of techniques are:

- Rotation: By keeping its origin at an angle (such as 90° or 180°), rotation can train an object recognition model for orientation invariance, thus enabling the recognition of objects in the input dataset regardless of their orientations. This is essential when objects in real-world scenarios present at all different angles—from medical imaging to autonomous driving. Rotating data for augmentation thus

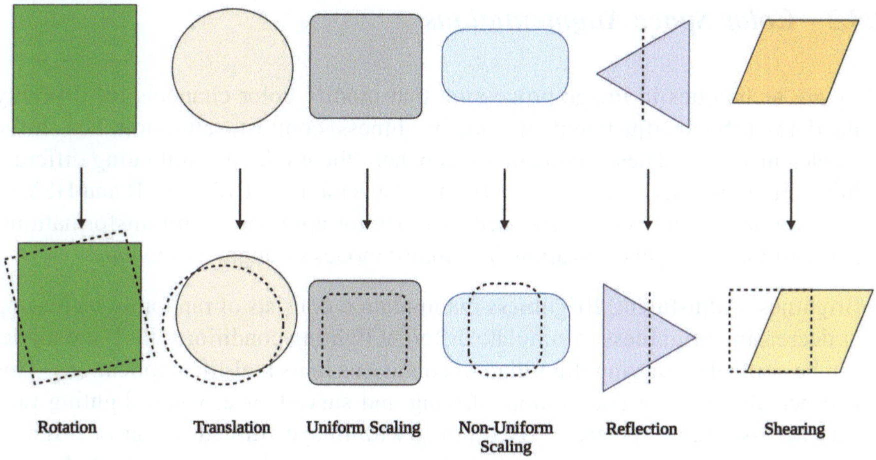

Fig. 2.3 Illustration of common geometric transformations. Examples include rotation, translation, uniform scaling, non-uniform scaling, reflection, and shearing, each showing the original shape (solid) and its transformed counterpart (dashed outline)

allows the model to become robust with respect to viewpoint variations, reducing overfitting and enhancing generalization [26].

- Flipping: The horizontal and vertical flips offered constitute very simple transformations but are meaningful to augmentations in algorithms for object detection tasks. This twin transformation provides a level of generalization to the models as they simulate mirrored views, calling for invariance to object orientation. For instance, flipping is extensively applied to datasets with symmetric objects like faces or vehicles. Text or directional features might be problematic candidates, as flipping might represent implausible variations. Nonetheless, flipping surely works as an inexpensive and powerful weapon for data augmentation [28].
- Scaling: Random zooming enhances recognition by training the model to perceive objects at varying scales, thus boosting its performance against size variance in real-world scenes. The prime application is medical imaging, where lesions are imaged at different magnifications, while in autonomous driving, size variance in distance objects is experienced. Thus, the lesson learned by providing a range of scales is a more robust feature representation that lessens sensitivity towards size variation [25].
- Cropping and Padding: Random cropping followed by resizing leads the model to part features and gains some robustness toward occlusion or variable object scales; padding helps keep the spatial dimensions while adding in some variability to avoid excessive dependence on the precise image borders. This is especially beneficial while trying to detect/classify different objects that may appear at varied positions or sizes in a real-life scene, thereby enhancing the generalization ability of the models to diverse scenarios [14].

2.3.2 Color Space Augmentations

They are techniques in image processing that modify color channels for diversity in the dataset. Such adjustments include brightness, contrast, saturation, hue shifts, and color inversion. These augmentations benefit the model by simulating different lighting environments and color distortions. Conversions between RGB and HSV or LAB color spaces are commonly used methods for applying such transformations, which will facilitate generalization. The major modes of augmentation are:

- Brightness Adjustment: Brightness manipulation consists of randomly increasing or decreasing brightness to simulate different lighting conditions; thus, the model can be applied in varying illumination conditions in its real-life applications. This is especially true for autonomous driving and surveillance, where lighting may vary across a large spectrum. By training with images turned darker or brighter, the model learns to extract features independent of the intensity of the light, thus improving its robustness.
- Contrast Modification: Adjusting the contrast would allow the model to generalize on different image qualities. By varying the intensity contrast between light and dark regions, this simulates various lighting conditions and camera settings, increasing robustness in real-life applications. Higher contrast augmentations target edge detection, while lower contrast variations force the model to rely on subtle features and avoid over-reliance on extreme pixel values. In medical imaging and low-light photography, such extremes of contrast become important.
- Hue and Saturation Shifts: Adjustments in hue and saturation are practical where color variations are a concern in the dataset and would simulate different lighting conditions and camera settings for better model robustness. Hue alters the color spectrum while saturation influences color intensity. This technique is beneficial for autonomous driving (varying weather and lighting) and medical imaging (staining variations). Such augmentations prevent overfitting to specific color distributions and improve potential generalization (e.g., medical imaging).
- Grayscale Conversion: Converting an image into grayscale reduces computational complexity while keeping structural features. Since color channels are eliminated, the model analyzes single-channel data, which increases training speed and lowers memory requirements. Grayscale conversion is useful in applications where color information is of less importance, such as texture analysis or edge detection. However, it may not work well for applications that require defects requiring discrimination based on color, such as object recognition in colored environments [24].

2.3.3 Kernel Filters and Noise Injection

Kernel Filters are kernel functions (as used in SVMs or GPs) applied to the data in order to smoothen or enhance features. Noise Injection is one form of regularization, which merely adds random noise to either inputs or weights during training, to gain robustness and discourage model overfitting. Both these functions augment the generalization of the machine learning models. Applying filters and noise introduces variation that can help in robustness:

- Gaussian Blur: Smoothing images by blurring kernels will allow the model to take care of out-of-focus data. High-frequency noise, simulating an image that appears out-of-focus either due to motion or less-than-perfect camera focus, can be reduced by applying a Gaussian filter on an input image. This technique ensures robustness whereby the model passes the performance test even when faced with imperfect inputs. In addition, controlled blurring is a form of regularization to avoid over-reliance on close details and to generalize in various image quality.
- Sharpness: Improvement in the visibility of features by sharpening makes edges and other features more easy to distinguish from each other, hence increasing model performance in applications associated with fine-grained feature detections such as medical imaging or object recognition. By applying convolutional filters that emphasize high-frequency components, sharpening overcomes blur and improves details in structure. Excessive sharpening introduces noise, and therefore, must be applied judiciously. According to several studies, combining sharpening with other augmentations, for example, Gaussian blur, enhances robustness.
- Random Noise: Adding Gaussian or salt-and-pepper noises forces the model to learn noise-invariant features improving robustness against the real-world scenario, where data might not be perfect. This reduces overfitting, encouraging the focus of the network on meaningful patterns instead of memorizing input details. Similar to that of sensor error or low quality input, noise injection can also act in making the model more elastic. Varying the intensity of noise is also intended to match the severity of regularization, allowing for generalization without loss of critical information [18].

2.3.4 Mixing Images

Image mixing is the process of merging two or more images into a single composite. Overlaying the images, changing their opacity, masking, and digital editing are some techniques used for this purpose. The applications of image mixing are in photography, graphic design, and art- enhancement of visuals, expression of themes, and some surreal effects. Advanced augmentation techniques involve merging multiple images:

- CutMix: A patch from another image replaces a slice from one image enhancing localization by smoothly juxtaposing the texture and colors for a natural look. This mode of augmentation is applicable in editing images, removing objects, and augmenting the data for machine learning. The features are aligned and lighting is adjusted to ensure coherence with the target image. Advanced methods use deep learning to fine-tune edges and increase their realism to a degree, rendering the modification visually undetectable while preserving context [27].

- MixUp: Generate two images with a linear mixture that creates a smoother decision boundary because it combines pixel values on weighted averages. It improves the visual transitions between pictures and reduces jaggedness in image morphing, alpha blending, and gradient-based effects. Users control the input of the two images seamlessly by varying the weights for blending. Such approaches find increasing their use in computer vision, photo editing, and deep learning to improve generalization in the model through imitation on the rather slow pace of feature transitions.
- Cutout: Randomly masking parts of an image forces a model to learn about diverse features and not to overfit specific characteristics. It helps robustness via modeling a masked image as it helps a network to infer from context to reconstruct regions which are absent. It improves generalization by changing the shapes and sizes of the masks, thus improving performance in object detection and segmentation. It also increases attention to detailed fine-grained features as well as to their global structures. Such approaches are pharmaceutically applicable in self-supervised learning, which uses unlabeled data in pretraining before fine-tuning in downstream tasks [22].

2.3.5 Adversarial and Neural-Based Augmentations

These are methods used for enhancing machine learning models. Adversarial augmentations refer to imperceptible perturbations in data to make it robust, while neural-based augmentations utilize synthetic training samples generated from generative models (like GANs or VAEs). Both these methods contribute towards adding more diversity and generalizing the datasets to involve better performances concerning the tasks they need to accomplish such as image classification. Recently there have been generative models into the newest advancements for augmentation:

- Generative Adversarial Networks (GANs): Produce fake images that imitate real distributions from data through the fighting of, say, two neural networks—a generator and a discriminator—against each other. These would consequently create fake images while the latter evaluates whether or not they are real. By means of iteration, it if eventually successful in training would make the GANs data which are now so realistic that could find applications in art and medicine and in augmenting data. However, mode collapse and training instability still exist. Improvements have kept showing through advanced forms such as DCGAN and StyleGAN. These variants improve stability and quality and reiterate GANs' critical importance in deep learning research and the synthetic media generation [13].
- Neural Style Transfer: Has been used to fuse unlike images of content and styles and generate fresh training samples to further enrich dataset diversity. With the help

of convolutional neural networks (CNNs), NST extracts the structural features of a content image together with the artistic patterns of the style reference and fuses them into a new output. It is clear that this technique would not only increase the generalization of the model but also minimize the effects of overfitting and use for many purposes, one of which could be found in art generation, medical imaging, and data augmentations for small datasets. NST, on the other hand, offers cool functionalities such as photo stylization, and its adaptive nature enables fine-tuning for certain domains as those involving fashion or simulations in autonomous drives [23].

2.4 Preprocessing for Object Detection

Preprocessing is a crucial requirement of any object detection pipeline, in fact it is the first stage for an object detection pipeline. Images and annotations need to be preprocessed before any deep learning model can take them to enhance accuracy and efficiency. Proper preprocessing can completely change the performance of object detectors by assuring uniformity, noise reduction and finally, optimizing input space for learning. Techniques included in this procedure are more than just normal image manipulation and include a variety of the following operations: Normalization, resizing, annotation refinement, manipulations for background, format conversion, and data cleaning techniques.

The section elaborates on the resourceful preprocessing steps which inarguably need to precede the application of object detection models, emphasizing how important these steps are to the performance of the model, their contribution to the training stability, in addition to their influence in resource consumption.

2.4.1 Normalization and Standardization of the Image

Normalization and standardization were important because they made sure that all images had similar distributions in pixel values. Therefore, it is important for the stabilization in the training of gradient descent algorithms and makes the model converge faster.

- Normalization is a process that usually scales pixel values within a specific range. Generally, this can be [0, 1] or [–1, 1].
- Standardized images have the form 0 mean and unit variance which can be achieved by using the formula Eq. (2.1):

$$X_{\text{standardized}} = \frac{X - \mu}{\sigma} \tag{2.1}$$

One of the things that transformations can do is lessen the effects wage by different lighting conditions and sensor characteristics [29]. A number of pretrained models such as YOLO, Faster R-CNN, and SSD expect that input images are normalized onto a given distribution defined on the training data [19].

2.4.2 *Image Resizing and Aspect Ratio Preservation*

One input that an object detection model usually expects as input is of fixed size regardless of the source. Resizing the image with a constant aspect ratio prevents the distortion of objects within the image that could otherwise impair performance with observations from the detection algorithms. Figure 2.4 shows the comparison of data normalization and standarization for the images.

- An empty area from the resizing may be filled using padding to maintain the respective original aspect ratios.
- Letterboxing, a method for padding, applies this approach by padding the resized image with a fixed value (typically black or gray) and is applied in YOLO models.

Incorrect resizing may cause elongation or squashing of the object appearance, influencing both anchor box matching and bounding box regression [46].

2.4.3 *Bounding Box Format Conversion and Sanity Checking*

Bounding box format conversion is to ensure compatibility with a certain annotation standard. Sanity checks for validating annotations that prevent the inclusion of invalid boxes (for example: with negative dimensions, out-of-image coordinates)

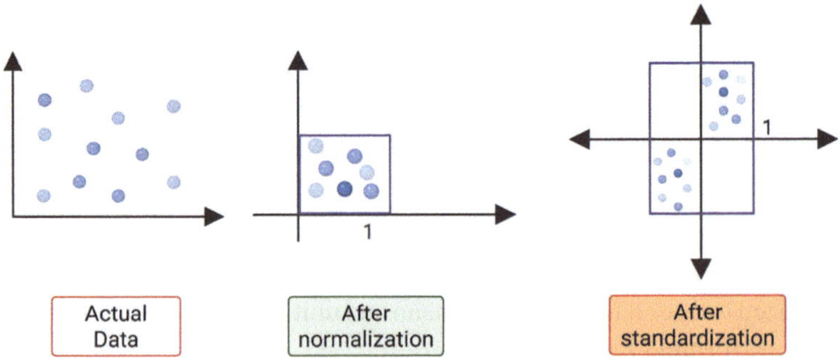

Fig. 2.4 Comparison of data normalization and standardization. The scatterplots show the original data distribution, the effect of normalization (scaling values to a [0, 1] range), and standardization (transforming data to zero mean and unit variance)

that may lead to training errors. The checks could either be automated via tools such as OpenCV or custom scripts for better reliability of the dataset and this, therefore, would allow correct formatting and validation to be an asset for the model's performance in relation to accurate ground-truth representation. Common formats include:

- (x_min, y_min, x_max, y_max): corner-based
- (x_center, y_center, width, height): center-based
- YOLO format: Normalized center coordinates and dimensions with respect to image size.

Before actual training begins, it is crucial to:

- Convert all boxes to one format
- Validate that bounding boxes do not go beyond image dimensions
- Validate all boxes for possible invalid entries (say, zero-area boxes, or negative coordinate).

Poor quality or inconsistent annotation can grossly affect model performance and sometimes lead to training crashes [12].

2.4.4 Background Cleaning and Clutter Removal

In object detection, background cleaning and clutter removal include preprocessing steps to eliminate images that would detract focus or distract the learning model so that meaningful features become emphasized. Such techniques as masking, edge detection, and semantic segmentation help in isolating target objects while suppressing noise. Reducing false positives and improving feature extraction in cluttered scenes would thus work well with these preprocessing attributes, as is the case in medical imaging or autonomous driving. By preprocessing, one would be doing:

- Area exclusion by background deduction: i.e., area exclusion perpetrated through either thresholding or semantic segmentation
- Blurring or Gaussian blurring: or in layman's terms decreasing the prominence of the background at the expense of fore-ging objects.

Such techniques work great when domain actions are concentrated in medical imaging or industrial inspection, where the interest is of clearly separated objects (Litjen).

2.4.5 Histogram Equalization and Contrast Enhancement

Histogram equalization is a method that attempts to redistribute pixel intensities for better display contrast by making certain dark or washed-out areas more discernible.

By spreading the more commonly used intensity values, it accentuates object boundaries and structural details in poorly lit and low-contrast images. This preprocessing stage helps with target detection because it enhances the visibility of features without affecting the actual data. Adaptive histogram equalization (CLAHE) extends this concept by lessening the amplitude of noise in uniform regions.

- Histogram equalization redistributes pixel intensities so that they cover the entire range more uniformly.
- Contrast Limited Adaptive Histogram Equalization has been found to be very useful in emphasizing local contrast without exaggerating noise.

These techniques are useful for underexposed or overexposed images and in grayscale datasets.

2.4.6 Image Denoising and Sharpening

Image noise, which arises from sensor imperfections and non-linear compression, hides critical features and deteriorates the performance of object detection. Denoising techniques such as Gaussian blur and median filtering reduce noise while preserving edges, while sharpening enhances fine details by accentuating high-frequency components to ease feature extraction. It is essential to strike a balance between the two processes: an excess of denoising may cause the blurring of crucial structures while excessive sharpening accentuates the noise. A well-done preprocessing gives a cleaner input to the models and improves the outcome in the real world.

- Median filtering: Very good at removing salt-and-pepper noise.
- Bilateral filtering: Reduces noise while preserving edges.
- Edge sharpening: Helps object detectors localize more accurately by highlighting boundaries.

Denoising especially plays an important role with low-quality surveillance tapes or sensor data [3].

2.4.7 Annotation Balancing and Dataset Cleaning

Annotation balancing inserts processing techniques—public class oversampling or loss weighting—into various steps of a learning problem so that the model can learn properly with class balance. Moreover, if we apply some processing methods for removing the incorrect data samples with labels, we call this operation data cleaning. This will go a long way in increasing the reliability of data. All the above will ensure that the training of object detection models is steady and generalization occurs across all classes. Real-world datasets often suffer from:

- Class imbalance: Where some object classes dominate others
- Redundant or conflicting annotations: Multiple bounding boxes for the same object or overlapping boxes with incorrect class labels.

To control for this:

- Undersample dominant classes or oversample minority classes
- Remove duplicate or ambiguous annotations
- Employ data cleaning tools like CVAT, LabelMe, or Roboflow for quality control [18].

2.4.8 Format and Framework Compatibility

Different object detection frameworks require specific input formats for optimal performance. For instance, TensorFlow Object Detection API typically uses TFRecords, while PyTorch-based models like Faster R-CNN or YOLO prefer COCO or Pascal VOC annotations. Some frameworks, like Detectron2, support multiple formats including JSON-based COCO datasets. Compatibility thus pertains to the adjustment of the annotations, resizing of the images to the expected size, and correct structuring of the data pipeline to prevent mismatches or anomalies associated with preprocessing during training and inference:

- TensorFlow's TFRecord
- PyTorch's custom dataloaders or COCO format
- Darknet format for YOLO.

Preprocessing entails converting the dataset into a compatible format, with all metadata, labels, and image paths structured correctly. File structure misalignment or mismatching image-label pairs may create problems during training.

2.4.9 Object-Centric Cropping for Fine-Grained Learning

In instances of multiple objects appearing in one scene or image, cropping around an object and treating it as one training sample assists the process of fine-grained learning. This intensity isolates single objects and diminishes background noise such that the model can concentrate on the discriminating features. By making object-poised inputs cleaner, the method helps boost recognition accuracy for the detail classes, aiming thus at:

- Enhancing class separation
- Highlighting fine details
- Counteracting context-based biases.

This is relevant above all to object detection tasks for small or overlapping items, such as biomedical cell or microbe detection.

2.4.10 Domain Adaptation and Dataset Harmonization

Domain adaptation improves model performance when training on multiple datasets or deploying across different domains by minimizing distribution shifts between source and target data. Techniques include feature alignment, adversarial training, and self-training. Dataset harmonization standardizes variations (e.g., lighting, resolution) across datasets to reduce bias. Both approaches enhance generalization, particularly in medical imaging or autonomous driving, where data heterogeneity is common, preprocessing may involve:

- Color matching: Aligning color distributions across datasets
- Resolution matching: Ensuring similar image quality
- Style transfer: Adjusting image styles for domain alignment.

Domain adaptation ensures that the learned features remain useful across different test distributions and mitigates domain shift, which can otherwise result in performance degradation.

2.5 Synthetic Data Generation

Synthetic data generation is an emerging trend in machine learning (ML) and artificial intelligence (AI) resulting from an increasing demand for large, very diverse, and high-quality datasets. Synthetic data generally refers to artificially generated information that mimics the statistical and structural features of real data but is not actually copied from it. This technique is very important in overcoming the issues such as shortage of data, privacy invasion, and bias towards the training data used for AI model development. With the advent of deep learning and computer vision along with generative models, artificial data have emerged into an irreplaceable component of the industry, ranging from health care and finance to autonomous driving and cyber security [36]. Figure 2.5 illustrates the Workflow of an autoencoder for data reconstruction.

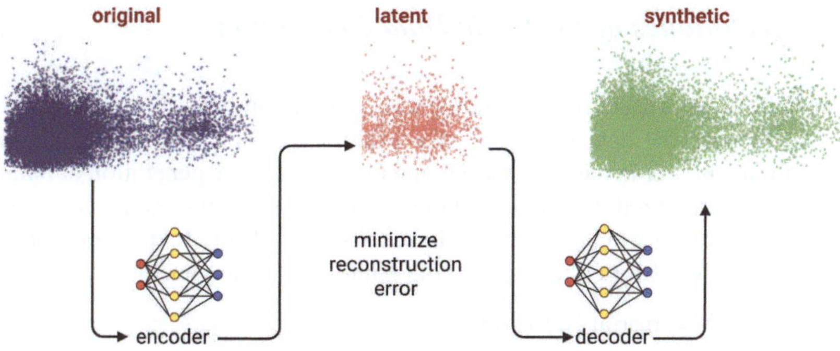

Fig. 2.5 Workflow of an autoencoder for data reconstruction. The original data (left, purple) is encoded into a lower-dimensional latent space (middle, red) using an encoder neural network. The decoder network then reconstructs the data in the synthetic space (right, green), with training aimed at minimizing the reconstruction error between original and synthetic distributions

2.5.1 Benefits of Synthetic Data

The advantages of synthetic data become very clear when real data is either:

- Scarce or very expensive to collect, such as rare disease images or very high-quality annotated video sequences, these challenges can make training very robust models extremely difficult because of limited availability and very high annotation costs. These realities strongly suggest the use of synthetic data generation, transfer learning, and few-shot learning to redress such deficiencies in training samples without compromising model accuracy and generalization.
- Under privacy laws and ethical concerns. Sensitive information, including patient health records or financial transactions, must be strictly anonymized and encrypted to avoid unauthorized access and comply with regulations such as HIPAA or GDPR. With proper precaution, confidentiality is maintained for individuals, but data may still be used responsibly.
- Imbalanced or biased so that it performs poorly on underrepresented classes or in the real world where the data distribution differs. This is especially problematic in applications like medical diagnosis or autonomous driving, where minority classes require accurate detection (e.g., rare diseases or edge-case obstacles). Models will make skewed predictions if there is no mitigation (such as resampling, synthetic data generation, or loss reweighting), leading to reduced reliability in production environments.

Some of the main benefits of synthetic data include Controlled diversity: Ability to program data distributions in order to have a broad representation of the conditions. Infinite scalability: Unlike real data, synthetic datasets can be generated in virtually unlimited numbers. Synthetic data can avoid showing personal identification directly and can be useful to the GDPR or HIPAA [37].

2.5.2 Techniques of Synthetic Data Generation

A multitude of approaches for synthetic data generation exist, but these approach vary widely with the data type used (for example, tabular, images, text) and the degree of realism. For instance, SMOTE and GANs serve the generation of balanced datasets from structured data. In the realm of images, diffusion methods and neural rendering techniques are adapted to enhance visual realism. Text generation relies on LLMs like GPT-4 for coherent and contextually relevant generation.

1. Generative Adversarial Networks:
   - GANs are the foremost among the tools to generate high-fidelity synthetic images, audio, and text.
   - A GAN consists of a generator and a discriminator in a zero-sum game, continually improving the output generated [13].
2. Variational Autoencoders:
   - VAEs learn the data distribution by encoding real samples into a latent space and decoding them back into synthetic samples.
   - However, for image generation, their output is less sharp than GANs, but they have stable training and are widely used for anomaly detection [35].
3. Rule-based and Simulation-based Methods:
   - Usually utilized for autonomous driving and robotics, where simulated data can reside in 3D environments such as CARLA or Unity.
   - These methods allow perfect control over object placement, lighting, motion, and environmental conditions [34].
4. Language Models for Text Generation:
   - The likes of GPT or BERT can be fine-tuned to generate domain-specific synthetic text for application in chatbot-oriented training, question-answering, or summarization [31].

2.5.3 Applications of Synthetic Data

Synthetic data is phenomenal and aids many domains such as:

- Healthcare: Anonymizing patient protocols wherein models can be trained with synthetic medical images (like MRI, X-ray) without compromising the identity of the patient. The MedGAN demonstrated such realistic medical image synthesis [33].

- Autonomous Vehicles: Simulators have generated a simulated driving environment that produces a synthetic sensor, i.e., radar, LiDAR, camera, which can be further used to train perception models.
- Cyberspace: Synthetic logs and attack paths are used to train intrusion detection systems without exposing any sensitive data related to an underlying network [38].
- Financial Reports: Generated synthetic credit card transactions allow a model to be trained as concrete and non-intrusive in terms of fraud detection [32].

2.5.4 Challenges and Limitations

Thus it is plain that, despite the promise, synthetic data generation brings forth yet another bag of challenges:

- Quality Assurance: Poorly generated data could contain artifacts or unrealistic patterns that may hamper model generalization, therefore introducing a certain hazard.
- Overfitting to synthetic patterns: Models which train on synthetic data exclusively would be ineffective when it is time for the real-world data transfer.
- Validation challenges: The verification process is not trivial at all due to the absence of thorough metrics and field tests when comparing synthetic data with real data.
- Bias amplification: If the generative model learns from biased real data, it can transmit and even accentuate these biases [39].

2.5.5 Evaluation Metrics for Synthetic Data

To prove the usefulness of synthetic data, a wide-ranging set of quantitative and qualitative metrics are employed:

- The Inception Score (IS) and Fréchet Inception Distance (FID) are designed specifically to capture how real the images are. IS evaluates both quality and diversity of images by employing a pre-trained Inception-v3 network, while FID asserts that the feature distributions of real and synthetic images provide a sounder measure of visual fidelity. An FID of 0 would therefore imply complete resemblance to the real, while an FID in the thousand's indicates synthesis of life-like images. Higher values of IS denote even better diversity and clarity of images. Hence, both metrics play a vital role in benchmarking GANs and other image synthesis-based models that ensure the generation of high-quality realist output. While IS emphasizes unnatural images with excessive sharpness, FID tends to measure perceptual similarity. Both assessment methods are employed so that a coherent and unified evaluation of the generative process can be obtained.
- Distribution similarity reviews such as Kolmogorov-Smirnov tests for tabular data are intended to assess whether two datasets follow the same statistical distribution

and are thus useful to validate synthetic tabular data or to detect dataset shifts. Such non-parametric tests compare two empirical cumulative distribution functions (CDFs) to gauge discrepancies to establish a level of agreement regarding feature engineering and preprocessing pipelines.
- Downstream performance: how well models trained on synthetic data generalize to the real world. If synthetic data misses diversity or realism, model accuracy could drop significantly in practical applications. Thus, high-quality data generation methods are needed.
- Human evaluation: domain experts evaluate the realism and quality of what was generated. This becomes especially critical in fields like medical imaging and natural language generation, where accuracy and comprehensibility are essential. This method ensures the synthetic data meets acceptable thresholds for realism, clinical utility, and contextual fit for training or use. The expert review will typically identify problems of a subtlety that might elude automated metrics.

2.6 Handling Imbalanced Data

Imbalances in datasets represent a common and challenging threat in actual ML applications, especially in domains such as medical and diagnostic problems, detection of fraud, and so forth, rare-event predictions [45]. According to the definition, imbalance occurs when the number of instances in one class very much outweighs any other classes, causing biased models that perform poorly on minority classes. Solving this issue is crucial in ensuring predictive performance that is comparatively robust and fair. Conventional ML algorithms usually work toward optimizing overall accuracy, assuming a balanced class distribution. However, in this scenario of imbalanced data, a model may flaunt high accuracy just predicting the majority class and neglecting the minority. This leads to:

- Poor recall and precision for the minority class
- High false negative rates, especially detrimental in critical situations like cancer detection
- Biasing of evaluation metrics, notably accuracy.

2.6.1 Handling Imbalanced Data-Methods

There are several approaches to addressing imbalanced data, which can be divided into data approach, algorithmic approaches, and hybrid methods.

- **Data-Level Methods**: These methods aim to modify the training data so as to achieve a balance in class distributions: one of the common methods is oversampling the minority class, which means there is an increase in the size of the underrepresented class. The oversimplest method is random oversampling, in which random duplications of the minority class samples are made to match the majority class size. Being simple, it has the dangerous possibility that the method may cause overfitting by reinforcing noise or outliers [41]. The better-performing algorithm employs SMOTE (Synthetic Minority Oversampling Technique), which generates new synthetic samples by interpolating between existing minority class examples. SMOTE currently enjoys wide acceptance because it enhances the diversity of the minority class, which may lead to improved generalization by the model. On the other hand, the vast majority of current methods work on reducing representatives of the majority class. This includes random undersampling, which entails removing random representatives from the majority class so that the class imbalance may be minimized. In essence, this method may drop important instances, thus affecting the performance of the model. More advanced methods such as Tomek Links or Edited Nearest Neighbors (ENN) mainly focus on removing ambiguous or borderline samples, helping to sharpen the decision boundary between the classes and increasing accuracy in classification [40]. Different strategies are used in practice combining oversampling and undersampling hybrids, playing both methods against each other. For example, applying SMOTE together with Tomek Links generates synthetic minority samples and at the same time cleans the majority class by removing overlapping or noisy data points. Overall, the hybrid method proves to work better than either SMOTE or Tomek Links alone [44].

- **Algorithm-Level Methods**: which assigns higher misclassification costs to the minority class. This approach ensures that misclassifying minority instances carries a greater penalty, thereby encouraging the model to focus on learning from these harder-to-classify examples. It is particularly effective in decision trees, support vector machines (SVMs), and deep learning frameworks [43]. Another widely used method is class weighting, where certain machine learning libraries—scikit-learn and XGBoost—provide built-in functionality for automatic balancing scheme adjustment of the importance of classes by assigning appropriate weight. This allows the model to pay more attention to underrepresented classes during system training. The ensemble learning techniques also work in favor of imbalanced datasets. An example is Balanced Random Forests, which modifies bootstrap sampling to ensure a balance in class representation for each decision tree [42]. Likewise, the EasyEnsemble and BalanceCascade methods generate multiple balanced subsets from the training data, training a separate classifier on each of them, and aggregating the results subsequently. This boosts the overall predictive performance, especially in cases of strong data bias [46]. These algorithmic strategies supplant the data-level alternatives convincingly, especially when preservation of the original distributions is of utmost importance or when resampling is impractical.

- **Deep Learning Approaches**: The performance of Deep Neural Networks is, however, considerably deteriorated due to imbalanced datasets, as they are more prone to overfitting the dominant class and underperforming on the minority categories. Several strategies have been proposed to resolve this. Among them is the highly effective strategy of modifying the loss function to put more weight on hard-to-classify or underrepresented samples. For example, Focal Loss has been proposed to give an adaptive scaling of the loss contributed by easy examples to focus learning on hard-to-classify examples and minority-class samples [30]. Another alternative is employing a class-balanced loss that adjusts the loss function based on the effective number of samples per class and, hence, ensures greater influence of rare classes over general optimization. Data augmentation can also be implemented to synthetically increase the size and diversity of the minority class. This is especially advantageous in fields like medical imaging or voice recognition, where the costs and efforts required to put a large annotated dataset together may be simply too high. Finally, transfer learning can act as a potent remedy whereby a model is pretrained on large balanced datasets and then fine-tuned on the imbalanced target dataset. This fine-tuning often improves the generalization ability of the pretrained model, thereby enhancing its chances for better recognition of minority class instances due to the prior knowledge stored in the weights.

2.6.2 Best Practice for Handling Imbalanced Data

Machine learning models can be drastically unbalanced when it comes to imbalanced datasets, causing issues with generalization of performance, especially with underrepresented classes. Thus, systematic best practices—from analysis all the way through to evaluation and model selection—should be adopted to promote well-balanced performance and fairness across all classes. Following measures outline a few rudimentary steps for effective handling of class imbalance:

- Analyze class distribution before modelling: An important first step is to understand the proportion of each class in your dataset. This helps to assess the degree of the imbalance and thereby guide the selection of techniques for handling it—for example, resampling or class-weight adjustment.
- Use visualization tools like confusion matrices: Confusion matrices work well to expose model misclassification errors, especially if the minority class is being neglected. Confusion matrices give a better picture than accuracy by itself.
- Avoid being overly reliant on accuracy: In the case of imbalanced data, an accuracy figure may be misleading. A model predicting just the majority class might achieve high accuracy, but would completely miss minority ones.
- Be willing to try several methods: Try many variations for handling imbalances; oversampling techniques such as SMOTE, undersampling, ensemble models, or cost-sensitive learning. The best method will greatly depend on the field of application.

- Cross-validate your performance: Stratified k-fold is a really great thing to use, to ensure that the distribution of classes in the original data is respected in all folds—greatly ensuring a fair performance evaluation.
- Watch for changes in model behavior: Observe how the model prediction changes with the application of balancing methods. This will help in understanding whether the gain in performance is not posed at the cost of overfitting the minority class.
- Utilizing domain knowledge: An expert can reveal possible avenues that could aid error analysis and show those types of misclassification that constitute the most important errors, particularly for healthcare and security.
- Reproducibility needs to be ensured: Provide sufficient information about data preprocessing, balancing methods, hyperparameter settings, etc., to maintain transparency and ensure results reproducibility.
- Iterate and fine-tune: Tackling the problem of imbalanced datasets rarely actually works after one single attempt. Iteration and ongoing evaluation are important to preserve the practicality of its performance on a skewed dataset of real-world inspirations.

References

1. Boutell, T.: PNG (Portable Network Graphics) specification. Network Graphics Group (2022)
2. Freeman, M.: The RAW file handbook. Focal Press, London (2023)
3. Gonzalez, R.C., Woods, R.E.: Digital Image Processing, 4th edn. Pearson, London (2018)
4. Google: WebP compression study. Google Developers (2021)
5. LeCun, Y., Bengio, Y., Hinton, G.: Deep learning. Nature **521**(7553), 436–444 (2022)
6. Murray, J.D., vanRyper, W.: Encyclopedia of Graphics File Formats, 2nd edn. O'Reilly, Sebastopol (2020)
7. Oppenheim, A.V., Schafer, R.W., Buck, J.R.: Discrete-Time Signal Processing, 3rd edn. Pearson, London (2021)
8. Prince, S.J.D.: Computer Vision: Models, Learning, and Inference. Cambridge University Press, Cambridge (2023)
9. Szeliski, R.: Computer Vision: Algorithms and Applications, 2nd edn. Springer, Cham (2022)
10. Wallace, G.K.: The JPEG still picture compression standard. IEEE Trans. Consum. Electron. **38**(1), xviii–xxxiv (2019)
11. Dutta, A., Zisserman, A.: The VIA annotation software for images, audio and video. In: Proc. ACM Int. Conf. Multimedia (2019). https://doi.org/10.1145/3343031.3350535
12. Everingham, M., Van Gool, L., Williams, C.K.I., Winn, J., Zisserman, A.: The Pascal Visual Object Classes (VOC) Challenge. Int. J. Comput. Vis. **88**(2), 303–338 (2010). https://doi.org/10.1007/s11263-009-0275-4
13. Goodfellow, I., Pouget-Abadie, J., Mirza, M., Xu, B., Warde-Farley, D., Ozair, S., Courville, A., Bengio, Y.: Generative adversarial nets. In: Adv. Neural Inf. Process. Syst. 27 (2014)
14. He, K., Gkioxari, G., Dollár, P., Girshick, R.: Mask R-CNN. In: Proc. IEEE Int. Conf. Comput. Vis. (ICCV) (2017). https://doi.org/10.1109/ICCV.2017.322
15. tzutalin2015labelimg, title=LabelImg, author=Tzutalin, Darrenl, journal=GitHub repository, volume=6, pages=4, year=2015
16. Intel: CVAT: Computer Vision Annotation Tool (2021). https://github.com/openvinotoolkit/cvat

17. Wang, Jason, and Luis Perez. "The effectiveness of data augmentation in image classification using deep learning." Convolutional Neural Networks Vis. Recognit 11.2017 (2017): 1–8
18. Lin, T.Y., Maire, M., Belongie, S., Hays, J., Perona, P., Ramanan, D., Dollár, P., Zitnick, C.L.: Microsoft COCO: Common objects in context. In: Eur. Conf. Comput. Vis. (ECCV). Lecture Notes in Computer Science, vol. 8693, pp. 740–755. Springer, Cham (2014). https://doi.org/10.1007/978-3-319-10602-1
19. Redmon, J., Divvala, S., Girshick, R., Farhadi, A.: You only look once: Unified, real-time object detection. In: Proc. IEEE Conf. Comput. Vis. Pattern Recognit. (CVPR) (2016). https://doi.org/10.1109/CVPR.2016.91
20. Settles, B.: Active learning literature survey. Univ. Wisconsin-Madison, Computer Sciences Technical Report 1648 (2009)
21. Long, J., Shelhamer, E., Darrell, T.: Fully convolutional networks for semantic segmentation. In: Proc. IEEE Conf. Comput. Vis. Pattern Recognit. (CVPR) (2015)
22. DeVries, T., Taylor, G.W.: Improved regularization of convolutional neural networks with Cutout. arXiv preprint arXiv:1708.04552 (2017)
23. Gatys, L.A., Ecker, A.S., Bethge, M.: Image style transfer using convolutional neural networks. In: Proc. IEEE Conf. Comput. Vis. Pattern Recognit. (CVPR) (2016)
24. Shorten, C., Khoshgoftaar, T.M.: A survey on image data augmentation for deep learning. J. Big Data **6**(1), 60 (2019)
25. Simard, P.Y., Steinkraus, D., Platt, J.C.: Best practices for convolutional neural networks applied to visual document analysis. In: Int. Conf. Document Analysis Recognit. (ICDAR) (2003)
26. Wong, S.C., Gatt, A., Stamatescu, V., McDonnell, M.D.: Understanding data augmentation for classification: When to warp? In: Int. Conf. Digital Signal Process. (ICDSP) (2016)
27. Yun, S., et al.: CutMix: Regularization strategy to train strong classifiers with localizable features
28. Krizhevsky, A., et al.: ImageNet classification with deep convolutional neural networks. In: Adv. Neural Inf. Process. Syst. (NeurIPS), pp. 1097–1105 (2012)
29. LeCun, Y., et al.: Efficient backprop. In: Neural Networks: Tricks of the Trade, pp. 9–48. Springer, Berlin (2012)
30. Lin, T.Y., et al.: Focal loss for dense object detection. In: Proc. IEEE Int. Conf. Comput. Vis. (ICCV), pp. 2980–2988 (2017)
31. Brown, T.B., Mann, B., Ryder, N., Subbiah, M., Kaplan, J., Dhariwal, P., et al.: Language models are few-shot learners. In: Adv. Neural Inf. Process. Syst. 33, 1877–1901 (2020)
32. Chen, X., Beutel, A., Xu, J., Chen, H., Chi, E.: Synthetic data in machine learning for credit card fraud detection. arXiv preprint arXiv:1901.10597 (2019)
33. Choi, E., Biswal, S., Malin, B., Duke, J., Stewart, W.F., Sun, J.: Generating multi-label discrete patient records using generative adversarial networks. arXiv preprint arXiv:1703.06490 (2017)
34. Dosovitskiy, A., Ros, G., Codevilla, F., Lopez, A., Koltun, V.: CARLA: An open urban driving simulator. In: Proc. 1st Annual Conf. Robot Learning, pp. 1–16 (2017)
35. Kingma, D.P., Welling, M.: Auto-encoding variational Bayes. arXiv preprint arXiv:1312.6114 (2013)
36. Nikolenko, S.I.: Synthetic Data for Deep Learning. Springer, Cham (2021)
37. Patki, N., Wedge, R., Veeramachaneni, K.: The synthetic data vault. In: IEEE Int. Conf. Data Sci. Adv. Anal. (DSAA), pp. 399–410 (2016)
38. Ring, M., Wunderlich, S., Scheuring, D., Landes, D., Hotho, A.: A survey of network-based intrusion detection data sets. Comput. Secur. **86**, 147–167 (2019)
39. Sandoval, R., Cruces, S., Balle, B.: The effects of bias amplification in synthetic data generation. In: Proc. NeurIPS Workshop on Dataset Curation and Security (2020)
40. Batista, G.E.A.P.A., Prati, R.C., Monard, M.C.: A study of the behavior of several methods for balancing machine learning training data. ACM SIGKDD Explor. Newsl. **6**(1), 20–29 (2004)
41. Chawla, N.V., Bowyer, K.W., Hall, L.O., Kegelmeyer, W.P.: SMOTE: Synthetic minority oversampling technique. J. Artif. Intell. Res. **16**, 321–357 (2002)
42. Chen, C., Liaw, A., Breiman, L.: Using random forest to learn imbalanced data. Univ. California, Berkeley, Tech. Rep. 110, 1–12 (2004)

43. Elkan, C.: The foundations of cost-sensitive learning. In: Proc. 17th Int. Joint Conf. Artif. Intell. (IJCAI), pp. 973–978 (2001)
44. Fernández, A., Garcia, S., Galar, M., Prati, R.C., Krawczyk, B., Herrera, F.: Learning from Imbalanced Data Sets. Springer Briefs in Machine Learning. Springer, Cham (2018)
45. He, H., Garcia, E.A.: Learning from imbalanced data. IEEE Trans. Knowl. Data Eng. **21**(9), 1263–1284 (2009)
46. Liu, X.Y., Wu, J., Zhou, Z.H.: Exploratory undersampling for class-imbalance learning. IEEE Trans. Syst. Man Cybern. B Cybern. **39**(2), 539–550 (2009)

Chapter 3
Object Detection Models

Abstract In this chapter, we explore algorithms used in object detection pipelines, both single-stage and two-stage architectures. We begin with an overview of single-stage detectors such as Single Shot Multibox Detector (SSD) and EfficientDet, as they employ a single pass for localization and classification processes to prioritize fast inference, as the ideal application of edge devices for real-time applications. After reviewing single-stage architectures, we then focus on two-stage architecture, specifically, the R-CNN family of detectors, which take precedence over accuracy by creating region proposals to localize objects before classifying each object, and at the price of inference time. We will then cover some specific architectural components that lead towards and advance the two-stage architecture, such as region proposal networks (RPNs), as significant to Faster R-CNN, and feature pyramid networks (FPNs), as part of Mask R-CNN, to demonstrate improvements in various tasks such as multi-scale object detection and instance segmentation. Finally, we will only briefly mention anchor-point free models, such as CenterNet, that anticipate that in most cases, one point can represent that object, thereby stripping down detection pipelines. The paper examines how speed, accuracy, and computation efficiency involve good trade-offs, and highlights where certain models are better suited for some application, e.g. autonomous driving vs medical imaging. At the end of the paper, the authors review deployment frameworks like TensorFlow Lite, which allow the optimization and use of these complex models on lower resource production hardware, thus allowing a means of getting academia and technological investment into usable/practical solutions for the real-world.

Object detection algorithms are the backbone of many intelligent systems that enable machines to perceive, comprehend, and act upon visual evidence in an online fashion. They are responsible for locating and categorizing instances of objects within an image or video stream, thus distinguishing themselves from image classification, which merely identifies objects. Object detection is the integration of both localization and recognition in a single process. This makes it quite a challenging and computationally intensive task. At an appropriate level of detail, object detection

A. Elhanashi and S. Saponara, *Deep Learning for Object Detection and Localization*,
https://doi.org/10.1007/978-981-95-4846-0_3

involves extracting important visual features, determining object boundaries, and associating the features with predefined object categories [19]. The traditional methods for object detection were mostly based upon hand-crafted features and statistical methods. For example, edge detection, color histograms, or gradient-based descriptors would be applied with region proposal methods to find candidate locations of the objects [5]. Unfortunately, these methods have limitations with variations in illumination, orientation, occlusion, or scale. Therefore, the accuracy and robustness of the early detectors were quite limited in applications where image capture and presentation vary so much with respect to quality and appearance of objects in real life [3]. The basic steps in an object detection algorithm are feature extraction, region selection, classification, and localization. Feature extraction involves finding low- or high-level descriptors that encode structural or semantic information about regions in the image. Region selection proceeds by eliminating large areas that are safely irrelevant and thus reduce computational overhead by focusing on possibly relevant segments. This classification involves assigning a score or probability to each region with regard to whether it holds the presence of the particular object class. Then localization infers the position as well as the size of each object, often defined in terms of locations of bounding boxes [9]. The newest methods for object detection go beyond manual feature engineering into fully automated feature learning to eliminate these deficiencies. The object detection algorithms therefore can learn the subject features from raw pixel information and adapt to variabilities present in visual scenes. Learned features are hierarchically organized, with the lower layers capturing edges and textures, while the higher layers capture complex shapes and object parts [11]. Such a hierarchical framework allows for greater resiliency for detection against occlusions, rotations, and varying object scales. Multi-scale detection plays an important role in boosting accuracy, especially when targeting small objects or those appearing at different zooms. Algorithms generally take inputs on various spatial dimensions while collecting contextual information from both global and local features. This concept permits an algorithm to obtain precise detection of objects in different sizes with high recall and precision [4]. Additionally, context modeling is applied to enable detection of clutters in scenes. The ability to model the spatial cues from different objects further enhances detection under the condition of ambivalence or interference [13].

Post-processing with NMS is yet another strong ally for object detection algorithms. In NMS, scores are assigned to each prediction, and while any remaining detections of the object(s) that receive a lower score are discarded, the one surveillance detection that receives the highest score is allowed to remain. Thus no multiple detections are produced for one instance, further ensuring accuracy in the final output [2]. More advanced versions of NMS implement other considerations such as class-wise suppression and density of detections. These methods further the performance of object detection algorithms due to data augmentation and transfer learning. Data Augmentation such as flipping, scaling, and cropping helps to build a more diverse sample thereby increasing the ability of the algorithm to generalize from training data into unseen scenarios [12]. On the other hand, transfer learning is the process in which an algorithm pre-trained on a larger dataset is then be fine-tuned on a smaller domain-specific dataset to reduce training times and improve performance in specific applications, such as medical imaging, surveillance, or autonomous navigation [6]. Real-time object detection in video poses additional challenges, such as temporal coherence and motion blur. For fast-paced applications including robotics and autonomous driving, algorithms must trade off speed with accuracy. Temporal smoothing and motion tracking strategies are embedded into the detection pipeline to ensure object identity across frames and diminish flickering effects [14]. This approach enhances the continuity and reliability of detections in dynamic environments. The evaluation of object detection algorithms is based on metrics such as precision, recall, and mean average precision (mAP), which are standardized. Such metrics assess a fair quantification of the algorithm's performance across classes and scenarios to enable results to be compared and benchmarked [1]. The continuous development of benchmark datasets and challenges also provides a stimulating cooperative environment for the research community which, in turn, leads to improved methodologies in object detection.

3.1 Single-Stage Object Detectors

These object detectors belong to the single-stage family and combine the two activities of localizing and classifying an object in a single go to become integrated as a stream-lined object detection method. Such a kind of scheme is totally unlike those methods taking part in separating these activities into different phases. It optimizes not only speed but also the simplicity of operation, rendering it most appropriate for real-time applications or deployment on edge devices with limited resources. The end-to-end architecture is designed such that it could operate on raw input images and produce object bounding boxes and class predictions in a single pass, as well as removing intermediate region proposal steps.

Realization of single stage detectors are motivated by the requirement of low-latency inference in real-time applications. To cite some common examples, they form part of applications such as autonomous driving, robotics, video surveillance, and mobile device applications. In such cases, speed is more important than accuracy, and one-cycle computation efficiency is especially beneficial [16]. This performance computation cost balance makes it popular. Figure 3.1 shows the simplified architecture of a single-stage detector.

1. Key Features

   The most defining features of single stage object detectors:

   - End-to-end pipeline: The whole task of detection is performed in a single network with lower complexity.
   - Grid-based prediction: Input images are actually divided into a grid, with each of its cells putting forth prediction of bounding boxes and class scores.

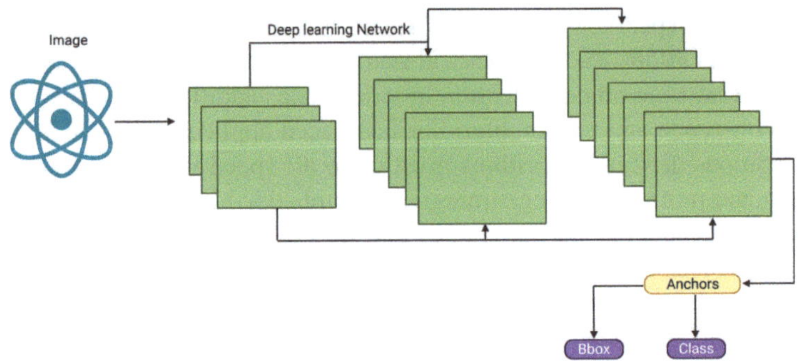

Fig. 3.1 Simplified architecture of a single-stage detector. The model uses a deep network and a set of predefined anchors to directly predict bounding box adjustments (Bbox) and class labels for each anchor, without a separate region proposal step

- Higher inference speed: The single pass architecture gives faster prediction times, which are suited for real-time systems.
- Scalability: Efficient memory and computation usage makes them fit for almost any hardware platform.

Though benefits these, single-stage approaches suffer some challenges in terms of localization precisions and detection accuracies, particularly with small or overlapped objects. Because of the object classification and bounding box regression done by the same network, errors in one would be propagated and would probably affect the other. Besides, there is the solution searched into loss functions and sampling techniques to alleviate this underperformance which might have caused by existing imbalance between background and foreground objects into the training data [19].

2. Training Strategies and Enhancements

A variety of training approaches have been put into the practice for improving single-stage object detectors, such as data augmentation, multi-scale training, hard-negative mining, focal loss optimization, etc. All these training strategies were mainly developed to overcome some fundamental problems like class imbalance, scale variance, and overfitting.

- Anchor-based designs give the detection framework an internal set of bounding boxes with varied combinations of aspect ratio/scale, providing prior hypothesis for object localization. Then, during training, the model learns to adjust these anchors for a better match with the ground truth objects. In this way, the network becomes fairly relative regarding objects of different sizes and shapes. Nevertheless, the major issue with anchor-based methods lies in the difficulty of determining the best fitting anchor to each ground truth box. This anchor assignment decides almost all convergence and precision with regard to the model. A mismatching anchor assignment will compromise detection performance, especially in dense or cluttered scenes where many objects overlap.

- Feature pyramid representations improve the capability of the model to recognize objects in images formed at different scales. These represent that More features are extracted at different depths in convolutional neural network and fused at top-down pyramid method. Functionally, quite shallow layer features serve to capture very high-resolution spatial levels that are useful for small object detection. Deep layer features encode rather high-level semantic features that are useful for recognizing bigger objects. This multi-level fusion of features helps the model to build a more wooden and consistent perspective of object properties regarding scales; hence, the strength and accuracy in detection greatly improve. Dramatic difference in sizes in actual images for objects depending on the distances in the camera proves the utility of such multiscale feature extraction.
- Single stage detectors already take great advantage of contextual modeling and attention mechanism. Contextual modeling is about using information from the immediate surroundings of a target object to make better decisions about its classification and localization. As an example, it could probably help recognize a traffic signal if he also sees the pole associated with it or street signs. Attention mechanisms go beyond this by emphasizing the very portions of the image that are relevant during inference such that the model may ignore unwanted noises in the background and prioritize features critical to recognition performance. This selective focus leads to a better model on object recognition, especially in complex scenes where multiple objects or distractors are present.

All these have continued to work towards shrinking the performance gaps with multi-stage systems and other computationally intensive approaches. These address enhanced characteristic representations, spatial sensitivity, and treatment of relatively heterogeneous object characteristics, while still retaining advantage in efficiency by single-stage architectures, significantly enhancing their operational usefulness across real-time surveillance applications, mobile computing, and autonomous systems [17].

3. Applications and Suitability

Half-dead short-only stage object detectors are in immediate need for all such applications in which inference time matters the most. A low-latency and computationally efficient design, which becomes simple, enables energy installation for all the different real-world applications on needing really fast, real-time decision-making. Such detectors demonstrated their effectiveness in the following well-known areas:

- Mobile systems and embedded devices: As with all mobile computing environments, including mobile devices, wearables, and even microcontroller-based on package domestics, there are certain limitations that were most primarily depended on by memory, power, and processing capabilities. Such were highly dependent when working without any accesses to a cloud server for inference on-device, which is important to avoid latencies while keeping the data private with low bandwidth consumption. Of course, having lightweight architecture allows for highly resource efficiency making it useful in providing a responsive behavior using even basic device configurations.

- Autonomous systems: Real-time perception of surroundings is important to move autonomously in a fully autonomous mode, without human intervention. A case would be a self-driving car, a drone that does deliveries, and a robotic assistant. Such applications require object detection for obstacle avoidance, pedestrian detection, or path planning. Delays in responses in such applications can pose grave safety threats. The way single-stage detectors can rapidly, reliably localize, and classify objects caters for instant decision-making in such dynamic environments, which might call for split-second decisions.
- Surveillance and monitoring: Smart cities, industrial facilities, and public infrastructures have video routines that relay on their smart video monitoring systems to watch their feeds in real time for possible such events as they happen: intrusion, unattended objects, or unauthorized access. It also scales in that, with the stream of video data, an advantage that single-stage detectors have over the ability to make such streams efficient in so many directions without incurring vast computational costs would be multiplied with the sheer volume of repetitive video analysis needed. Multiple deployments across multiple camera feeds in parallel are enabled because of the high throughput capacity of these detectors; therefore they can cover large areas and be much more responsive.
- Augmented reality (AR): AR-based applications deal with superimposing virtually created elements into a user's naturally existing environment with a sense of real-time immersion. Accordingly, this involves the accurate and real-time detection and tracking of real-world objects in a manner pertinent to the ability of tracking objects in real-time with low latency. Although the applications may be gaming, navigation, industrial maintenance, or interactive education, real low-latency and accurate object detection merge these activities into a smooth AR experience. Single-stage detectors partly satisfy this requirement through consistent yet very low-latency object acknowledgment required for more interaction and user interface responsiveness.

The deployment-friendly availability on edge devices is strongly aided by the development of quantized and pruned deep learning models that reduce size and computation demand with only modest accuracy trade-offs; quantization is the process of shrinking parameter precision (e.g. from 32-bit floating-point to 8-bit integers), while pruning removes redundant or less important weights in the network. Such technologies have permitted the implementation of single-stage detectors on constrained hardware, such as microcontrollers and IoT platforms, and have thereby potentially increased their applicability in environments where energy consumption and responsiveness act as key factors [15].

4. Limitations and Research Directions

Although these light methods might be efficient in such a way that latency is low sometimes, the same difference is in accuracy when compared with a complex method, especially with fine-grained recognition, occlusions, and dense scenes. One major difference occurs when comparing their performance with that on small or overlapping objects, where fine-contextual knowledge is mostly required. To address these challenges, research is aimed at improving algorithmic components without adding loads to computation. Adaptive loss functions develop learning to focus on hard-to-learn cases for imbalanced data, while dynamic assignment techniques help to make better localizations while at the same time reducing mismatches during processing. These techniques all add up to quite a few more improvements but not speed. Memory usage and processing efficiency sufficiency in today's resource-constrained environments such as IoT and TinyML becomes consulted vastly. Of course, removing restraints such as using quantization, pruning, or platform-specifications is made to fit somehow into microcontrollers and low-power edge devices. They may therefore see their way into real application systems such as wearables, smart cameras, and autonomous sensors. It is certainly here to stay. Increasingly gaining adoption, semi-supervised and self-supervised learning methods can compel the use of huge volumes of unlabeled data, hence, educating raw inputs in pseudo-labeling and contrastive learning without the need for massive manual annotation to induce better generalization. Transfer learning and knowledge distillation are important in embedding previously learned knowledge into smaller systems that could perform efficiently via compression and fine-tuning. There is fast growth in interest in the development of transparency and interpretability issues to enable trustworthy operation, especially in sensitive applications. Overall, it comprises research that is still ongoing relative to the improvement of efficiency against accuracy, ensuring small, responsive systems remain reliable and appropriate for a wide variety of real-world applications [18].

3.1.1 You Only Look Once (YOLO) Model Series

The YOLO series—which stands for You Only Look Once, refers to real-time object detection models that combine speed and efficiency. Incepted in 2016 by Redmon [7] YOLOv1, the architecture has gone through multiple iterations of the model, yielding a lot of refining in terms of performance, robustness, and efficiency of deployment. By 2024, the family has matured to a total of 11 iterations YOLOv11, where every iteration has concentrated on a specified resolution of problems faced during detection tasks, such as speed versus accuracy, small object detection, and adaptation to edge devices. A single convolutional neural network (CNN) [7] was used by YOLOv1 to enforce a unified architecture, framing object detection as a regression problem defined by predicting bounding box coordinates and class probabilities directly from whole images. The benefits were fast detection, but it was poor on small-object detection and localization errors. YOLOv2 or YOLO9000 improved on this model through batch normalization, high-resolution classifiers, and anchor boxes. Multi-scale training has been employed. The detection performance was then extended to over 9000 categories by joint training on the classification and detection datasets [8]. Another feature that was introduced by YOLOv3 is the improvement of the object detection pipeline using more powerful backbone networks like Darknet-53, multi-scale predictions through feature pyramid networks (FPN), objectness prediction with logistic regression, and multi-label classification effect using independently trained binary

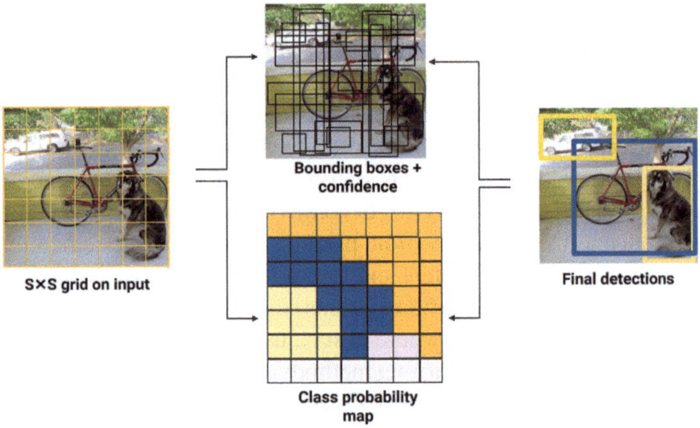

Fig. 3.2 Illustration of a single-stage detection approach (e.g., YOLO). The input image is divided into an S×S grid, where each grid cell simultaneously predicts bounding boxes, confidence scores, and class probabilities to generate final detections

classifiers to boost the performance on detecting small and overlapping object [9]. YOLOv4 [10] is an emerging name in its own accord as it signified a giant leap ahead in coupling the Bag of Freebies with the Bag of Specials to get the greatest boost possible in accuracy and speed without increasing inference costs. Backboned on CSPDarknet53, Mish activation, SPP, and PANet were employed for optimized feature extraction and fusion. Hence, it became quite stable and widespread because of the trade-off between speed and accuracy at standard GPUs. Figure 3.2 shows the illustration of a single-stage detection approach (YOLO). The Ultralytics-released YOLOv5 was the first YOLO model implemented in PyTorch, thus promoting its accessibility to research and industry. The initial criticism was mostly regarding the versioning since Ultralytics was not the original creator; however, YOLOv5 gained momentum because of its modular design with ease in deployment, auto-learning of anchor boxes, availability of different pre-trained models of various sizes (s, m, l, x) fit for much broader applications. Constructed by Meituan with industrial performance in mind, YOLOv6 has a decoupled head design and anchor-free detections. That gave YOLO6 incredible speed and mAP on the COCO benchmark which are really high numbers when referring to throughput. YOLO7, which was put forth by [32], breaks new ground in the simultaneous application of speed and accuracy by introducing the E-ELAN Extended Efficient Layer Aggregation Network-compound model scaling strategy and a new backbone for object detection. Outperforming the state-of-the-art in real-time object detection, YOLOv7 is said to be the future of this domain. Widely considered the future of computer vision, YOLOv8 reassured anchor-free generation by supporting segmentation and classification tasks and delivering CSPNet as a new backbone. Improvements have been made by YOLOv8 with respect to training equipment and export for deployment on platforms including TensorRT, ONNX, CoreML, and TFLite. YOLO9 onwards kept enhancing lightweight shipments, hardware adaptability, and edge performance. A model of YOLO9 was introduced that implemented neural architecture search (NAS), transformer-enhanced attention modules, and improved quantization-aware training (QAT) techniques to significantly compress the model while incurring minimum loss in accuracy. YOLO10 concentrated more on transformer-CNN hybrid backbones and adaptive inference pipelines that boast dynamic computational scaling based on complexity of input. It greatly enhanced real-time segmentation and tracking, thus broadening YOLO's usability in autonomous driving and medical imaging. Designed around multi-resolution feature aggregation, lightweight transformer heads, and novel loss hybridization of CIoU with focal loss to address class imbalance and improve box localization, YOLO11 goes further in its applications. The support for federated training setups opens applications in privacy-conscious settings such as healthcare and smart surveillance systems. Table 3.1 shows a comparative Overview of the YOLO (You Only Look Once) Object Detection Model Evolution.

3.1.2 Single Shot Multibox Detector (SSD)

SSD is a well-known deep learning-based object detection framework with high speed and accuracy at the same time. Thus, it becomes one of the most important frameworks for real-time applications. SSD was proposed by [24], and since then one of the top single-stage detectors in the computer vision domain has outperformed two-phase detectors such as Faster R-CNN in accuracy while it can beat their time of inference. The SSD framework consists of a single deep neural network to detect multiple objects in the image and predicts object categories and bounding boxes in a single pass across the network, which is referred to as a single shot. This is opposed to a two-stage detector like the R-CNN [22], where the first step involves the generation of region proposals, followed by the classification of target objects. Both tasks are to be performed in a unified pipeline by SSD under this approach. This greatly improves the computational performance efficiency.

SSD is basically a network, normally one of the VGG-16 versions [25], which serves as a feature extractor. More convolutional layers are raised upon the base network to allow for the decreasing size of the feature map all the way to many scales of detection. This multi-scale approach, where SSD can effectively detect smaller and larger objects, is one of its key advantages. Each position in these feature maps is linked to a set of fixed default bounding boxes (also known as anchor boxes) of different aspect ratios and scales. Default boxes do all the job of locating the shapes and sizes of different objects efficiently in the SSD. The model predicts both the offsets with respect to these default boxes and the confidence scores for the presence of object categories in those boxes. A significant advantage of SSD is that it utilizes multiple feature maps across different layers in order to do object detection. The different spatial resolution feature maps thus act as a scale indicator for recognizing objects going small to large. Fine-grained features captured by early layers are used

Table 3.1 A comparative overview of the YOLO (You Only Look Once) object detection model evolution

Version	Backbone	Detection type	Key innovations	Strengths
YOLOv1	Custom CNN	Single-stage	Unified detection, single CNN	High speed
YOLOv2	Darknet-19	Anchor-based	Anchor boxes, batch norm, multi-scale training	Better accuracy
YOLOv3	Darknet-53	Multi-scale	FPN, Residual blocks, Logistic classifiers	Improved small object detection
YOLOv4	CSPDarknet53	PANet + SPP	BoF, BoS, Mish, data augmentation	Accuracy and speed balance
YOLOv5	CSPDarknet	Anchor-based	PyTorch, Auto-anchor, Model size variants (s, m, l, x)	Modularity, Ease of use
YOLOv6	EfficientRep	Anchor-free	Decoupled heads, Post-processing optimizations	High-speed inference
YOLOv7	E-ELAN	Multi-scale	E-ELAN, model re-parameterization, compound scaling	State-of-the-art real-time performance
YOLOv8	C2f Backbone	Anchor-free	Unified detection/ classification/ segmentation, ONNX & TFLite export	Versatility, Deployability
YOLOv9	NAS + Transformer	Hybrid	NAS, Quantization-aware training, Transformer layers	Lightweight and edge-device friendly
YOLOv10	Hybrid CNN-Transformer	Adaptive	Adaptive inference, Federated learning support	Context-aware detection
YOLOv11	Transformer-C2f	Multi-task	Enhanced multi-resolution attention, Federated training, New loss functions	Real-time + privacy-aware inference

for detecting smaller objects and are transformed into large abstract features reflecting more robust detection soon into the deeper layers.

To illustrate, SSD300-the most well-known of SSD variants-uses feature maps from layers of size 38×38, 19×19, 10×10, 5×5, 3×3, and 1×1, as they have all made contributions to the detection pipeline. This key diversity in the receptive fields and spatial resolutions has boosted SSD's capability to generalize across varying object sizes without the need for a separate region proposal network. SSD specifies a number of default boxes for every cell within a feature map based on different scales and aspect ratios. During the training phase, every ground truth object is matched to that default box with the highest overlap using the Jaccard overlap (Intersection over Union, IoU). This matching process allows kissing these positive matches primarily in case the overlap threshold with IoU is greater than 0.5; otherwise, they are treated as a negative match. This matching mechanism makes it possible for SSD to handle multiple object categories and densities in a single image while keeping the computational overhead low. Each matched default box provides an offset to be predicted from the center of the box along with classification scores. The SSD loss function has only two components: the first one is localization loss, and the second one is confidence loss. The localization losses are often smooth L1 losses between predicted box coordinates and those of the ground truth box, while the confidence losses are softmax losses over several classes that include a background class. The total loss is computed as Eq. (3.1):

$$L(x, c, l, g) = \frac{1}{N} (L_{\text{conf}}(x, c) + \alpha \cdot L_{\text{loc}}(x, l, g)) \tag{3.1}$$

where:

- L_{conf} is the confidence loss.
- L_{loc} is the localization loss.
- x represents the matching indicator.
- c is the class label.
- l and g are the predicted and ground-truth bounding box parameters, respectively.
- α is a weighting factor (commonly set to 1).
- N is the number of matched default boxes.

1. Performance and Benchmarks
   SSD, in effect, has great App performance on standard benchmark datasets such as PASCAL VOC and MS COCO. According to [24], an mAP on the PASCAL VOC 2007 test set of 74.3% is attained for SSD300 at 59 frames per second (FPS), thus putting the model in a great position compared with Faster R-CNN, which achieves rather higher accuracy with an extremely low frame rate. Another variant, SSD512, enlarges the input image for accuracy enhancement but increases the computational burden and lowers frame rates to 512×512 pixel images for small object recognition. These small trade-offs render it very utilitarian for different applications of SSD—from heavy-performance machines to embedded devices.

2. Applications and Deployments
   Some of the application domains that SSD can cover include:

   - Detection of pedestrian and vehicular movement, real-time surveillance through videos.
   - Autonomous vehicles, where latency is paramount to safety.
   - Augmenting reality applications on mobile platforms.
   - Medical imaging, where the slight support is in identifying the anomalies in the scanning radiological images.

However, given the condition of efficiency that defines SSD, it can also be deployed on resource-strapped edge devices, especially when coupled with lightweight base networks such as MobileNet [23]; this has borne much popularity for the TinyML applications of SSD, which require low-power inference.

3. Strengths and Limitations

Strengths:

- Real-time execution: The architecture allows fast inferences with minimal compromises in accuracy.
- Multi-scale detection: The SSD detection layers are designed to efficiently capture objects across a wide range of input scales.
- Allow end-to-end training: Being single-stage in nature, the architecture thus optimizes and deploys aids simplification.

Weaknesses:

- Detection of small objects: This may give SSD an underdog position in the comparison to two-stage detectors about the small object detection.
- Overfitting in dense scenes: Conversely, in the event that many small and overlapping objects are present in one image, the behavior of fixed default boxes can make SSD perform poorly.
- Fixed aspect ratios: A range of various default box shapes are used by SSD, which may still struggle to generalize when faced with highly irregular object shapes.

3.1.3 EfficientDet

In this EfficientDet, as developed by Tan et al. [31], the family of object detection models offers efficient aspects of model effectiveness and computational resource investment. It generally extends the EfficientNet architecture and combines it with the well-known compound scaling method and a superlative feature fusion technique called the Bi-directional Feature Pyramid Network (BiFPN). Bugler-wise EfficientDet models balance performance with computational efficiency, enhanced for

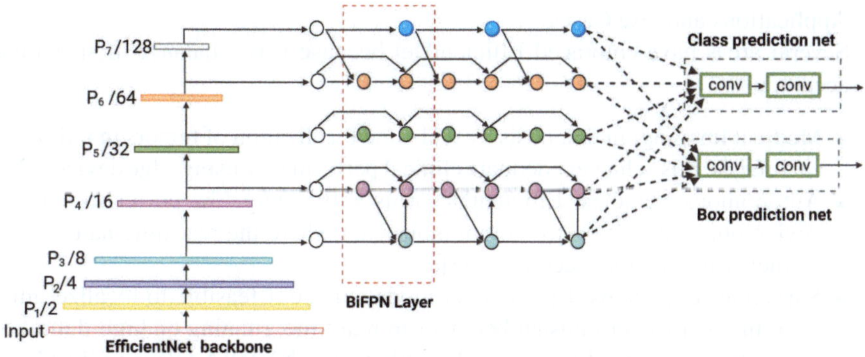

Fig. 3.3 Architecture of efficient single-stage detector (based on EfficientDet). It uses an EfficientNet backbone, a Bi-directional Feature Pyramid Network (BiFPN) for multi-scale feature fusion, and separate subnetworks for class and box prediction at different feature levels (P1-P7)

deployment in resource-constrained edge devices like mobile phones and embedded systems. Figure 3.3 illustrates the architecture of efficient single-stage detector (based on EfficientDet).

1. Innovations in Architecture
   The core innovations introduced by EfficientDet are the backbone network that is EfficientNet and the incorporation of BiFPN into the design for multi-scale feature fusion.
   EfficientNet Backbone: EfficientNet is a family of convolutional neural networks, advanced using a neural architecture search [30]. It adopts a compound scaling mode, scaling to provide uniform depth, width, and resolution while giving an extremely high accuracy-to-complexity ratio. Therefore, EfficientDet employs the lightweight and high accuracy of EfficientNet backbone for feature extraction from images, which has been proved to perform outstandingly on the classification tasks.
   BiFPN: The Bi-directional Feature Pyramid Network (BiFPN) replaces traditional FPNs (Feature Pyramid Networks) by allowing easy and fast multi-scale feature fusion with learnable weights. Unlike traditional FPNs that use fixed feature fusion strategies, BiFPN introduces bidirectional cross-scale connections and adds learnable weights to combine features more effectively [31]. This will allow the network to pay more attention to those features which contribute more towards improvement in detection accuracy.
   Compound Scaling for Object Detection: following compound scaling in EfficientNet, Improved-Det extends the method to object detection by uniformly scaling the resolution, depth, and width of the backbone with BiFPN concerning the class/box prediction networks [31]. Thus, this compound-scaling process led to the formation of a model family ranging from EfficientDet-D0 (lightweight) to EfficientDet-D7 (larger and more accurate), so that developers can select a model based on the resources they have and accuracy requirements.

2. Applications and Use Cases
   Several areas have embraced EfficientDet because it is a balance of speed and accuracy:

   - Medical Imaging: in diagnosis by real-time localization of tumors in radiological images; this achieves adequate clinical performance using edge devices [32].
   - Autonomous Vehicles: EfficientDet runs sufficiently low power to suit on-device object detection for autonomous cars where the real-time nature must be met with maximal accuracy [33].
   - Surveillance Systems: EfficientDet models make it feasible to localize intrusion, fire, smoke, or unusual behavior in real time, running on high-definition video streams on any Flea9 embedded platforms, NVIDIA Jetson or Raspberry Pi 4 with Coral Edge TPU [28].
   - Agricultural Automation: Plant diseases, weeds, fruits, etc., can be detected in a resource-constrained environment using effective deployment of EfficientDet models [29].

3. Deployment on Edge Devices
   EfficientDet has such really lightweight models, especially D0-D3, designed for edge computing. These models can then be quantized and pruned with several frameworks for running on microcontrollers, smartphones, and embedded systems. Examples are TensorFlow Lite and TensorRT. So far, it has been shown through experiments that EfficientDet-D0 and D1 could run in real-time of up to 20 FPS in Raspberry Pi 4 or NVIDIA Jetson Nano, optimized by 8-bit quantization. Deployment of EfficientDet has been applied to mobile hardware in video surveillance operations, detecting people, vehicles, and safety equipment with great accuracy [29]. Such extraordinary generalization brings additional robustness to the practical application.
4. Limitations and Challenges
   Constraints in EfficientDet applications are:

   - Training: Suffice it to say, the training of model counts high computational expense. It is particularly costly for higher versions like D6 and D7 as they require high to very high-end GPUs with large memory.
   - Quantization: EfficientDet accepts quantization but not aggressively post-training quantization, as it may decay the performance much worse while evaluating small objects and under-represented classes.
   - Domain Adaptability: Like many other CNN-based detectors, EfficientDet will require additional retraining in new, completely separated domains and environmental changes (lighting conditions, occlusions, etc.).

3.1.4 CenterNet

CenterNet, as defined [20], is one among a class of one-stage object detection architectures that completely reorients the entire object detection pipeline. The photorealistic object detection systems depend on localizing the key points of an object as its centers and relative distances from object center point for all properties associated with the object, like size, dimensions, and orientation. This approach eliminates anchor boxes and region proposals from traditional object detection architectures such as Faster R-CNN or SSD, leading to simple detection with comparable accuracy and speed.

1. Architectural Overview
   The basic idea of CenterNet relies on measure objects using predicting their geometric middle with the help of a heatmap. Once the center of an object is identified, it predicts its width and height as well as other possible parameters reference to the orientation of the object or 3D pose. The architecture normally comprises a backbone convolutional neural network (CNN) such as Hourglass, ResNet, or DLA (Deep Layer Aggregation) that extracts feature maps from the input image. These feature maps are then passed through specialized branches:

   - Center Heatmap Prediction: This is where a heatmap branch predicts center points of all objects in the image, having Gaussian peak in each object center.
   - Size Regression: This branch regresses the height and width of the bounding box associated with each detected center point.
   - Offset Prediction: This will determine location center by predicting sub-pixel offset.

   Thus, the approach converts the detection problem into pixel-wise prediction problem, exactly akin to semantic segmentation or point detection, which is going to increase computation efficiency and reduce inference time [34].

2. Major Contributions and Features
   Some of the major innovations in CenterNet are proposal-free and anchor-free designs, which take an enormous complexity and computation burden off the model. Unlike SSD or YOLO that require large sets of predefined anchor boxes for detecting objects, CenterNet throws away these heuristics and reduces memory usage as well as speeds up inference.
   Furthermore, CenterNet brings in household techniques such as those used in human pose estimation—from CornerNet and the Hourglass network experimental setup—which allows for accurate localization of object centers.
   Last but not least, center pooling and cascade corner pooling are introduced, two techniques that will improve the accuracy of center point localization. The pooling techniques collect with respect to horizontal and vertical directions, thus allowing the model to learn what the true centers are among cluttered or occluded scenes.

3. Variants and Extensions
   CenterNet has several variations, which support either an additional function or enhanced performance:

   - CenterNet-hg (Hourglass): It stacks Hourglass networks into a backbone for high accuracy at slow inference speed.
   - CenterNet-ResNet/DLA: These two use either ResNet-18/50 or DLA-34 where better trade-off performance in speed versus accuracy can be attained.
   - Object as Points (CenterNet): Extensions are made to the original model to study as 3D object detection and pose estimation in the autonomous driving settings through the prediction of additional attributes [34].
   - CTDet: Types of 2D object detection—these settings have high accuracy on benchmark datasets like COCO.

4. Applications
   Due to simplicity in design and effectiveness, CenterNet has flown beyond ordinary 2D-object detection:

   - Autonomous driving: CenterNet has been modified to identify and localize 3D bounding boxes for vehicles and pedestrians from monocular images [34].
   - Medical Imaging: Under the circumstances that demand precise localization of anatomical landmarks or lesions, Centers methodology works using keypoints.
   - Surveillance Systems: Low latency and a firm detections capacity are involved in providing real-time human detection and tracking.
   - Facial Landmark Detection: CenterNet has been used in face detection and keypoint estimation owing to its central focus.

5. Limitations and Challenges
   Even with its advantages, some of the limitations that CenterNet has include:

   - Occlusion Handling: A single-center representation is in the case of very severe occlusion or overlapping of objects where closely located centers are to be distinguished.
   - Object Size Variation: Objects that are very small or very large are still a challenging task and it becomes very difficult when the resolution becomes too low or too high concerning feature maps.
   - Multi-class Complexity: As the number of object classes increases and each of them becomes quite similar in features—such as the different types of vehicles—the increase in false positives is evident as the prediction of center points gets confused.

The majority of the problems posed by such issues have been tackled through the means of later models, which integrate multi-scale feature aggregation and improved data augmentation or attention mechanisms, that improve the capability of CenterNet to deal with dense or complex environments.

6. Recent Advancements and Impact
   CenterNet has greatly changed the face of object detection by creating an anchor-free, keypoint-based approach that simplifies detection pipeline and yet retains high accuracy. The design philosophy has motivated several subsequent detectors, which include FCOS (Fully Convolutional One-Stage Object Detector) and ExtremeNet, that share similar philosophies based on per-pixel predictions and localizing object centers instead of predefining anchor boxes. With this paradigm of simplicity and reduced computational costs, they are better suited for real-time and embedded applications. In parallel, among the transformer-based models, this paradigm added actually robust global attention mechanisms against the complexity of the detection tasks. Notably, with certain advantages, the aforementioned models may cause CenterNet to remain a dear entity for lightness and proficiency. This remains a vital choice in the context of actual deployments on edge devices and mobile platforms, where memory and computation constrictions call for small and fast models that do not sacrifice performance.

3.2 Two-Stage Object Detectors

Two-stage object detectors are, right now, quite popular in computer vision owing to their accuracy and solid performance in rather complex detection tasks. The two-stage concerned basically divide the task of object detection into two major components. Here, in the first stage, the algorithm aims to propose candidate regions in the image where objects are likely to be. Those proposed regions will then be forwarded to the second stage, where classification and localization will be further refined. This task separation means that a lot of computation can be applied to a few, select, promising regions, hence contributing to more accurate object detection. On the contrary, single-stage detectors try to perform classification and localization on the complete input image in a single shot and work against each other-the two-stage detectors decouple this process, giving the advantage to each step to sharpen its own view on a particular detection aspect. This contrast carries significance. The single-stage detectors are designed to be fast, recognizing objects on a dense grid of anchors directly, but often give up on accuracy if the objects are small or overlap each other. However, in the two-stage system, the first stage suppresses all background regions, allowing attention only on object-like regions in the second stage for detection, resulting in a better-performing detection system in the fine-grained localization or multi-class discrimination tasks [26]. The region proposal is hardly just one of the two pillars of two-stage detectors. The primary or first phase involves scanning through the incoming image for generating extremely few region proposals that contain a high probability of objects. These proposals are not undergoing classification, but might be served into certain regions being considered in detail. This path reduces the computational load on the second stage inherently, for only a small number of regions are being processed in depth. Thus, the two-stage architecture strikes a good computational cost-performance balance, well suited to applications where accuracy

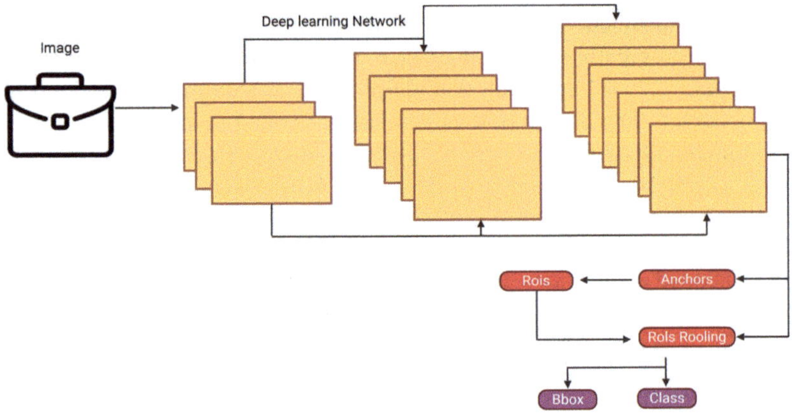

Fig. 3.4 architecture of a two-stage object detection model. The process involves generating Region Proposals (RoIs) from anchors, refining their coordinates (Bbox), and classifying the content within each region

is paramount, such as autonomous driving, surveillance, and medical image analysis [36]. Figure 3.4 shows the simplified architecture of a two-stage object detection model.

This is an intermediate step of the pipeline follow-up to the regional proposal stage; feature extraction was carried out on every possible region and then subjected to classification and bounding box regression. The class of the object to be targeted and the accuracy of coordinates around the identified objects is the target for the present phase. According to this view, the two-step refinement process becomes relatively more proficient in handling different scales and positions of the image and thus gives better precision and recall metrics with an overall better outcome than a one-stage counterpart [37]. In addition, this flexibility in architecture allows two-stage frameworks in having more modules added to increase their performance. For instance, it can also allow more seamless integration of context modeling with multi-scale feature fusion and post-processing strategies such as non-maximum suppression. Such enhancements are normal requirements to deal with occlusions, clutter backgrounds, or multiple object proximity challenges in the real world. The modular stage architecture of two-stage detection thus provides a very strong foundation for experimentation and development of advanced detection systems which promise a lot for researchers and practitioners through additional benefits from further study of such systems [39]. However, it is noted here that two-stage detectors are computationally slow compared to single-stage counterparts. This is naturally so as this course command takes a longer time and has subsequent stages for the precision checking of the outcomes generated. Hardware acceleration advances coupled with optimization technologies and model compressions have gradually alleviated these limitations such that in some applications, two-stage methodologies would now be approaching a near real-time performance [35]. Bargain between one stage and two stages

Table 3.2 Key Characteristics and Performance Trade-offs Between Single-Stage and Two-Stage Object Detectors

Feature	Single-stage detectors	Two-stage detectors
Architecture	Performs object classification and localization in a single step	Splits detection into two steps: region proposal and classification
Speed	Faster; suitable for real-time applications	Slower due to sequential processing
Accuracy	Generally lower than two-stage, especially for small or dense objects	Higher precision, especially in complex scenes
Complexity	Simpler and more streamlined architecture	More complex due to separate proposal and refinement stages
Computation cost	Lower computational cost	Higher computational cost
Training process	Easier and faster to train	More involved due to multi-stage design
Object scale sensitivity	Struggle with detecting small or overlapping objects	Handles small and overlapping objects more effectively
Localization precision	Less accurate bounding box localization	More accurate bounding box refinement

compromises speed against accuracy intrinsic in the meaning of object detection. Certainly, single-stage ones would be the preferred option of all real-time applications like robotics or mobile phones, while two-stage frameworks traditionally would be applied in those domains where the reliability and precision in detection aren't allowed to be compromised. The performances of these models are especially sought after in safety-critical areas since they can be used for their excellent performance in detecting small, distant, or partially occluded objects [38]. A comparison table summarizes the key differences between the two stages and a single stage: architecture, performance, and application. Table 3.2 illustrates the Characteristics and Performance Trade-offs Between Single-Stage and Two-Stage Object Detectors.

3.2.1 Region-Based Convolutional Neural Networks (R-CNN, Fast R-CNN, Faster R-CNN)

Region-based convolutional neural networks are the main forces of impact in this area and will be the future extensions, Fast R-CNN and Faster R-CNN. Using this architecture, we are of utmost importance in terms of how objects are identified and localized within an image, since it provides both more accurate and smoother computation than any other approach. This integrated model enables one to have a region proposal mechanism with the CNN and lays the foundation for state-of-the-art object detection systems.

The first R-CNN, which was launched by [22] in 2014, marks an important departure from previous pipelines based on object recognition by coupling region proposals with CNN-based feature extraction. Before the advent of R-CNN, object recognition was largely dependent on mechanisms such as the Deformable Part Model (DPM), which consisted of engineered features and exhaustive search strategies via sliding window paradigms. They usually had limited accuracy as well as computational efficiency. R-CNN revolutionized the process by combining selective search for region proposals with feature extraction based on deep learning [22]. The R-CNN pipeline begins with selective search to generate about 2000 category-independent region proposals from the input image, then warps each proposal into a fixed size (e.g., 224×224 pixels), and passes it through a pretrained CNN (such as AlexNet) to extract a feature vector. Finally, the feature vectors are entered into class-specific linear Support Vector Machines (SVMs) for classification, and a separate regression model is used to refine the bounding box predictions.

The application of R-CNN in real-time systems was hindered by some major limitations despite its better accuracy when compared to earlier methods. First, extracting features of each of around 2000 region proposals separately was computationally expensive, taking up to 47 s per image with the GPU. Second, the pipeline was not an end-to-end trainable model; it was comprised of three different stages: fine-tuning of the CNN for region classification, training of SVMs, and training bounding box regressors [22]. Therefore, the need for an improved model called Fast R-CNN was borne out. Fast R-CNN, which was introduced by Girshick [41], is one other significant improvement to the R-CNN architecture as it addressed the two main concerns of speed and training complexities. While R-CNN treats each region proposal separately, Fast R-CNN processes the entire image using a single pass through a convolutional network, producing a shared feature map for future use. As an alternative to cropping out the regions on the image, the RoI pooling layer extracts fixed-size feature vectors directly from the shared feature map corresponding to each corresponding region proposal. The feature volume is then fed to fully connected layers with two outputs: softmax classifier for object recognition and linear regression layer for bounding box prediction [41].

This change in architecture leads to and quite significantly causes different improvements in computational costs due to redundancy removal in the CNN computations. It allows, furthermore, end-to-end training using single-stage multi-task loss function, which optimally co-factors classification and localization. This mode of joint training gives a measurable performance improvement that is also less fair in implementation. Fast R-CNN was actually a fast, accurate competitor to earlier systems at standard benchmarks like PASCAL VOC and MS COCO. Still, it was premised on an external region proposal method, such as a selective search, which is computationally very burdensome and not learnable. So, practical real-time applicability was very limited.

To overcome this bottleneck of slow and not learnable region- proposal generation, It came up with Faster R-CNN in 2015 as a fully end-to-end, trainable model for object detection [45]. The end-to-end model of Faster R-CNN encapsulates a Region Proposal Network (RPN) into the pipeline of a CNN, thereby greatly eliminating the need for selective search. The RPN shares convolutional features to the detection network and generates object proposals by sliding a small network over the feature map. At each location, it simultaneously predicts objectness scores and regresses bounding boxes for multiple predefined anchor boxes of various scales and aspect ratios. Fast R-CNN, which was introduced by Girshick [41], is one other significant improvement to the R-CNN architecture as it addressed the two main concerns of speed and training complexities. While R-CNN treats each region proposal separately, Fast R-CNN processes the entire image using a single pass through a convolutional network, producing a shared feature map for future use. As an alternative to cropping out the regions on the image, the RoI pooling layer extracts fixed-size feature vectors directly from the shared feature map corresponding to each corresponding region proposal. The feature volume is then fed to fully connected layers with two outputs: softmax classifier for object recognition and linear regression layer for bounding box prediction.

This change in architecture leads to and quite significantly causes different improvements in computational costs due to redundancy removal in the CNN computations. It allows, furthermore, end-to-end training using single-stage multi-task loss function, which optimally co-factors classification and localization. This mode of joint training gives a measurable performance improvement that is also less fair in implementation. Fast R-CNN was actually a fast, accurate competitor to earlier systems at standard benchmarks like PASCAL VOC and MS COCO. Still, it was premised on an external region proposal method, such as a selective search, which is computationally very burdensome and not learnable. So, practical real-time applicability was very limited (Fig. 3.5). To overcome this bottleneck of slow and not learnable region- proposal generation, . It came up with Faster R-CNN in 2015 as a fully end-to-end, trainable model for object detection [45]. The end-to-end model of Faster R-CNN encapsulates a Region Proposal Network (RPN) into the pipeline of a CNN, thereby greatly eliminating the need for selective search. The RPN shares convolutional features to the detection network and generates object proposals by

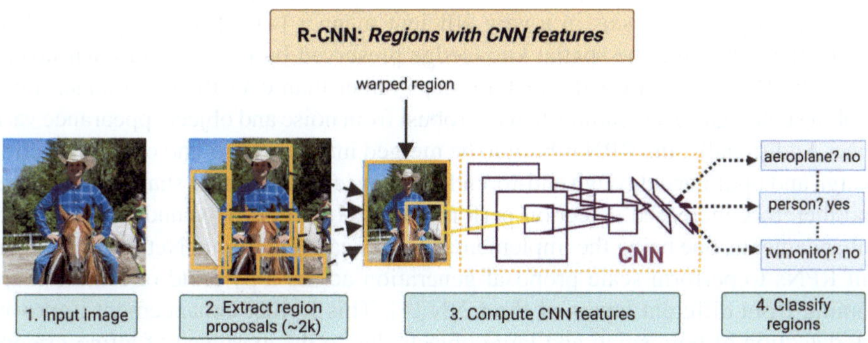

Fig. 3.5 R-CNN (Regions with CNN) pipeline. The model first extracts region proposals, warps them to a fixed size, computes CNN features for each region, and then uses a classifier (e.g., SVM) to determine the object class

sliding a small network over the feature map. At each location, it simultaneously predicts objectness scores and regresses bounding boxes for multiple predefined anchor boxes of various scales and aspect ratios.

3.2.2 Region Proposal Networks (RPNs)

Region Proposal Networks (RPNs) have become central features of recent models of object detection, mainly because they are the models used in two-stage detectors such as Faster R-CNN. RPNs are responsible for producing the candidate bounding boxes called region proposals, which can subsequently be refined and classified by further stages of the detection network. With the advent of RPNs, considerable evolution toward object detection has been achieved with region proposal generation being embedded within a neural network to make the process speedier and more accurate [45]. Before the advent of RPNs, region proposal generation was a computational bottleneck, where the traditional techniques like Selective Search or EdgeBoxes used hand-crafted algorithms to identify object-like blobs within an image and by using a CNN for classification [47, 48]. This approach was effective but not coupled with the learning process and, even worse, not designed for end-to-end training so could not adapt and optimize joint to that of the detection network. Instead, the RPNs can be defined in the straight-up convolutional manner having shared layers with the detection network, thus yielding a large yield in computation. For example, the RPN takes in a feature map being output from the backbone CNN (e.g., VGG-16, ResNet) and uses a small sliding window over it to generate anchor boxes with multiple scales and aspect ratios. An objectness score computed on each anchor box is then regressed, [45]. At each position on the feature map, there are k anchor boxes and the RPN outputs for every position 2k objectness scores (foreground or background) and 4k regression coordinates to refine the anchor boxes. The RPN is thus trained with multi-task loss combining objectness with bounding box regressions to refine anchor positions, such that it learns to produce excellent quality candidate region proposals in a very efficient manner [41]. The strength of RPNs comes from utilizing the spatial knowledge preserved by the convolutional layers. Since the RPN slides over the feature maps rather than over the raw image, it can exploit such high-level features that are robust from noise and object appearance variance. Additionally, the RPN is built to be merged into an end-to-end object detection design and optimized through end-to-end learning, enabling better sharing of features and therefore increased detection performance [51]. RPNs have undergone several enhancements, one being the implementation of Feature Pyramid Networks that permit RPNs to perform scale proposal generation across a pyramid of feature maps coming from different layers of the CNN [4]. This further enhancement improves the detection of both small and large objects due to the existence of a fine-grained information preservation mechanism across scales. Other methods target the improving of the input proposal quality. Region Proposal Networks utilizing focal loss [21] solve the class imbalance associated with object detection by minimizing the effect of well-classified examples in the loss computation, which forces the network to pay more attention to hard-to-classify examples. Other variants include anchor-free RPN methods, like FCOS [46], which predict object centers instead of having to rely on predefined anchor boxes, thereby simplifying architecture and enhancing generalization. More traditional object detection applications have found a near-endless deployment for RPNs. In medical imaging multiples of RPN-based frameworks have been employed to detect lesions or abnormalities with great precision in the radiograph [44]. They show that the architecture's generalizability and robustness can back their application in domain-specific object detection tasks. Table 3.3 shows the comparison of R-CNN, Fast R-CNN, and Faster R-CNN object detection architectures.

Table 3.3 Comparison of R-CNN, Fast R-CNN, and Faster R-CNN object detection architectures

Feature	R-CNN	Fast R-CNN	Faster R-CNN
Region proposal method	Selective search (external)	Selective search (external)	Region Proposal Network (RPN), integrated into CNN
CNN architecture	AlexNet (or other pre-trained CNNs)	Shared CNN for entire image	Shared CNN for entire image
Training phases	Three phases: CNN fine-tuning, SVM training, Bounding Box Regression	End-to-end training with a joint multi-task loss	End-to-end training with joint RPN and detection network
Speed	Slow (47 s/image on GPU)	Faster (2 s/image on GPU)	Faster (200ms/image on GPU)
End-to-End training	No (requires separate training for each module)	Yes (end-to-end with shared feature map)	Yes (end-to-end with RPN and detection network)
Region proposal	~2000 region proposals per image	~2000 region proposals per image	Region proposals generated by RPN (faster and integrated)
Accuracy	High accuracy but slow inference	Improved accuracy and faster than R-CNN	State-of-the-art accuracy with faster inference
Computational complexity	High (independent feature extraction for each proposal)	Lower than R-CNN (shared feature map)	Lower than Fast R-CNN (integrated RPN)
Object localization	Bounding box regression after SVM classification	Bounding box regression with softmax classification	Bounding box regression with RPN and softmax classification
Limitations	Slow, complex pipeline with high memory requirements	Still uses slow region proposal (selective search)	Still has latency due to two-stage architecture

Still, RPNs are not devoid of challenges. The main threat refers to dependency on a complaint of pre-fixed anchor shapes which might not generalize well to different object sizes and aspect ratios, spurring research in dynamic anchors generation and anchor-free models. Also, the performance of RPNs is sensitive relative to the quality of the underlying feature map which, therefore, imposes a very careful design on the backbone CNN and feature aggregation strategy [40].

3.2.3 Feature Pyramid Networks (FPNs) Mask R-CNN

FPNs have brought in a novel architectural advance in the area of object detection and instance segmentation, providing better feature representation across multiple scales. FPNs, which were introduced by [43], are already widely implemented as backbone in top-of-the-line deep learning models, such as Mask R-CNN. Its capability of capturing multi-scale contextual information has brought tremendous improvement in detection and segmentation tasks, especially on images with various sizes of objects.

- **FPN Architecture Outline**
  At the bottom of FPNs is the concept of constructing a pyramid of feature maps containing high semantic information at all levels. Usually, feature hierarchies in CNNs-res, such as those of VGGNet and ResNet-are naturally due to successive layers in the CNN, with the last few layers being semantically strong but low in spatial resolution. The FPN does the gaining of such features by connecting them through a top-down pathway and lateral connections to link them to high-resolution and semantically weak features [43]. This may be considered as a series of feature maps from high-reasoned to anatomically reasoned. The architecture typically consists of a bottom-up pathway, a top-down pathway, and lateral connections. The conventional bottom-up pathway is the forwards pass through the CNN which generates properties that progressively become more abstract. Inference through top-down pathway involves upsampling of the relatively coarse coarser but semantically stronger features. Lateral connections merge such upsampled features to similar high-res features from bottom-up pathway, thus augmenting both spatial accuracy and semantic richness.
- **FPNs in Mask-RCNN**
  Mask R-CNN, an advancement over Faster R-CNN, has FPN as an integral part to further enhance the feature extraction ability [42]. In Mask R-CNNs, the region proposal networks (RPNs) and the ROI (Region of Interest) heads work on the multiscale feature maps since FPNs have been used in these models. Consequently, the model is capable of sustaining strong performance on objects distributed over varying scales—a common difficulty in instance segmentation tasks.
  Also, FPNs ensure that Mask R-CNN will at least afford the opportunity to process every object with relevant semantics at a corresponding level of the feature pyramid. Thus, the RPN, working in tandem with appropriate heuristics, chooses to generate its proposals at multiple scales; given the size of each proposal, it is

then mapped onto the level of the pyramid such that the most suitable features for that scale attend to its processing. This scale-dependent processing helps the model localize and classify better. The same enhanced feature map is also used to compute the segmentation masks predicted per region, which further enhances the accuracy and detail of instance segmentation outputs.

- **Benefits of FPNs**
  The ability of FPNs in Mask R-CNN to deal with scale variation among objects is one of the biggest advantages. The same object may be portrayed with very different apparent sizes in regard to distance and/or intrinsic size variation in the real world. Traditional CNNs have very little chance of exhibiting any consistent performance across such scales. FPNs, while allowing for a balanced and rich description of the data, will enable the model to address both detection and segmentation of small and large objects [43]. At the same time, FPNs enhance computational efficiency. Instead of building separate feature extractors for various scales, a single backbone combined with FPN will treat all scales as one coherent problem. This single-framework style of operation thus improves speed and resource square-off productive, fitting for real-timers and large-scale applications [50].
- **Uses and Enhancement**
  In turn, FPN-integrated Mask R-CNNs have made their way into different fields such as medical imaging, automotive driving, and video surveillance. In tumor segmentation and organ delineation, FPN-based Mask R-CNNs have scored an advantage over traditional approaches [34]. In an automotive environment, multi-scale detection is thus required for pedestrians, vehicles, and distant road signs [49]. From the considered improvements on the original architecture of FPN to improve its performance, PANet introduces bottom-up path augmentation to allow better low-level feature propagation [52]. The same is true for NAS-FPN, which applies automated architecture searches to achieve optimal feature pyramids that outperform their manual counterparts [50].

3.3 TensorFlow Lite

In today's world, where AI and deep-learning models are used in everyday technology, deploying these models efficiently has become more and more important on edge devices. Traditional AI frameworks like TensorFlow can be considered powerful, but their computation- and memory-hungry nature often makes them unsuitable for embedded and mobile applications. To tackle this challenge, Google has initiated the launching of TensorFlow Lite (or TFLite), which is a lightweight way to deploy machine learning (ML) models onto devices that have very limited resources, such as smartphones, microcontrollers, and Internet of Things (IoT) devices. Herein, we will discuss the architecture, benefits, use cases, and current developments concerning TensorFlow Lite.

- **Architecture and Components of TensorFlow Lite**
  TensorFlow Lite performs the conversion of a full TensorFlow model into a smaller version that is optimized in a format called FlatBuffer file (.tflite), and this optimized format is efficiently interpretable on embedded systems. This process can be broadly divided into two phases: model conversion and model inference. In the conversion phase, optimizations including quantization, pruning, and operator fusion are being applied for transformation of a TensorFlow model using the TFLite Converter.
  The inference engine has an extremely lightweight runtime interpreter that works with a subset of TensorFlow operations optimized for edge devices. It can run on a variety of backends including CPUs, GPUs, and dedicated hardware accelerators like Edge TPU (Google Coral) and NNAPI (Android Neural Networks API), ensuring the flexibility on different platforms [53].
- **Primary Features and Benefits**

  1. Model Optimization: TFLite has full-fledged support for post-training quantization that minimizes precision in weights and activations changing the original 32-bit floating point to int8 representation, leading to a small model size, an improvement in performance and tolerable accuracy loss [57].
  2. Low Latency Small Footprint: Most TFLite models will use memory under 1MB, and their inference time can drop to only a few milliseconds, depending on the task and hardware used, rendering them best suited for real-time applications [55].
  3. Cross-platform Compatibility: TFLite is inclusive of support for Android and iOS, but it also supports embedded Linux via TensorFlow Lite for Microcontrollers (TFLM), which enables almost all possible types of cross-platform deployment.
  4. Hardware Accelerated: TFLite has also been able to define delegates that will offload parts of the computation to specialized chips like GPU, DSP, or TPU, which further increase performance [54].

- **Use Cases and Applications**
  Making it a highly sought-after trait in many domains because of its portability-performance efficiency, TensorFlow Lite has proven to be applicable in:

  – Health: Diagnostics are done through TFLite with real-timeity as would point the smartphone diagnosis.
  – Home automation and the IoT: Edge devices that use TFLite carry out speech recognition and gesture detection and perform home automation without connecting to the cloud [58].
  – Autonomous vehicles: Small models from TFLite are used for lightweight object detection models, which can be mounted onboard drones and small robots for avoidance of obstacles and navigation purpose [23].
  – Agriculture: Farmers in limbo internet countries use TFLite-based applications to diagnose diseases on their crops using their mobile phones [56].

- **TinyML and Microcontrollers**
  TinyML is indeed among the most striking integrations of TensorFlow Lite because it provides ML models in microcontrollers, with memory and storage sizes of kilobytes. TensorFlow Lite for Microcontrollers runs on architectures like ARM Cortex-M and RISC-V, thus, models as small as 20 KB can run on them. These systems find wide applications in voice-trigger systems, vibration sensors, and health-monitoring wearables [53]. TFLM is likely to be deployed in strongly embedded environments as it is independent of operating systems and it does not dynamically allocate memory. By going through the TensorFlow Lite Model Zoo, several pre-trained models such as speech keyword spotting and accelerometer-based gesture recognition become very easy for integration.
- **Challenges and Limitations**
  Nevertheless, TensorFlow Lite does have a few demerits, such as:

  1. Limited Operator Support: Many TensorFlow operations are left unsupported in TFLite; hence the developers have to alter the models or simplify them.
  2. There could be a lot of Conversions Error: conversion can fail for complex TensorFlow models or custom models requiring reconstruction into simple basic ones, which would cause extensive debugging or custom kernel development work needed.
  3. Performance versus accuracy trade-offs: quantization results in increased performance, but at the cost of reduced accuracy; hence it must be remained carefully tuned [57].
  4. Debugging on Edge Devices: The process of model debugging is inherently more difficult in resource-constrained systems due to limited tools and observability.

References

1. Everingham, M., Van Gool, L., Williams, C.K.I., Winn, J., Zisserman, A.: The PASCAL Visual Object Classes (VOC) Challenge. Int. J. Comput. Vis. **88**(2), 303–338 (2010)
2. Hosang, J., Benenson, R., Schiele, B.: Learning non-maximum suppression. In: Proc. IEEE Conf. Comput. Vis. Pattern Recognit. (CVPR), pp. 4507–4515 (2017)
3. Jiao, L., Zhang, F., Liu, F., Yang, S., Li, L., Feng, Z., Qu, R.: A survey of deep learning-based object detection. IEEE Access **7**, 128837–128868 (2019)
4. Lin, T.Y., Goyal, P., Girshick, R., He, K., Dollár, P.: Focal loss for dense object detection. In: Proc. IEEE Int. Conf. Comput. Vis. (ICCV), pp. 2980–2988 (2017)
5. Liu, L., Ouyang, W., Wang, X., Fieguth, P., Chen, J., Liu, X., Pietikäinen, M.: Deep learning for generic object detection: A survey. Int. J. Comput. Vis. **128**, 261–318 (2020)
6. Pan, S.J., Yang, Q.: A survey on transfer learning. IEEE Trans. Knowl. Data Eng. **22**(10), 1345–1359 (2010)
7. Redmon, Joseph, et al. "You only look once: Unified, real-time object detection." Proceedings of the IEEE conference on computer vision and pattern recognition. 2016
8. Redmon, Joseph, and Ali Farhadi. "YOLO9000: better, faster, stronger." Proceedings of the IEEE conference on computer vision and pattern recognition. 2017

9. Redmon, J., Farhadi, A.: YOLOv3: An incremental improvement. arXiv preprint arXiv:1804.02767 (2018)
10. Bochkovskiy, Alexey, Chien-Yao Wang, and Hong-Yuan Mark Liao. "Yolov4: Optimal speed and accuracy of object detection." arXiv preprint arXiv:2004.10934 (2020)
11. Ren, S., He, K., Girshick, R., Sun, J.: Faster R-CNN: Towards real-time object detection with region proposal networks. In: Adv. Neural Inf. Process. Syst. (NeurIPS), pp. 91–99 (2015)
12. Shorten, C., Khoshgoftaar, T.M.: A survey on image data augmentation for deep learning. J. Big Data **6**, 60 (2019)
13. Singh, B., Najibi, M., Davis, L.S.: SNIPER: Efficient multi-scale training. In: Adv. Neural Inf. Process. Syst. (NeurIPS), pp. 9310–9320 (2020)
14. Wang, Q., Wu, B., Zhu, P., Li, P., Zuo, W., Hu, Q.: ECA-Net: Efficient channel attention for deep convolutional neural networks. In: Proc. IEEE/CVF Conf. Comput. Vis. Pattern Recognit. (CVPR), pp. 11534–11542 (2021)
15. Howard, A.G., et al.: Searching for MobileNetV3. In: Proc. IEEE/CVF Int. Conf. Comput. Vis. (ICCV), pp. 1314–1324 (2019)
16. Liu, W., et al.: Object detection and recognition: Theory and practice. Pattern Recognit. Lett. **130**, 1–10 (2020)
17. Ren, S., et al.: Deep learning for generic object detection: A survey. Int. J. Comput. Vis. **129**(2), 441–518 (2021)
18. Sze, V., Chen, Y.H., Yang, T.J., Emer, J.S.: Efficient processing of deep neural networks: A tutorial and survey. Proc. IEEE **105**(12), 2295–2329 (2017)
19. Zhao, Z., et al.: Object detection with deep learning: A review. IEEE Trans. Neural Netw. Learn. Syst. **30**(11), 3212–3232 (2019)
20. Zhou, Qiang, et al. "Instant-teaching: An end-to-end semi-supervised object detection framework." Proceedings of the IEEE/CVF conference on computer vision and pattern recognition. 2021.
21. Lin, T.Y., et al.: Focal loss for dense object detection. IEEE Trans. Pattern Anal. Mach. Intell. **42**(2), 318–327 (2020)
22. Girshick, R., Donahue, J., Darrell, T., Malik, J.: Rich feature hierarchies for accurate object detection and semantic segmentation. In: Proc. IEEE Conf. Comput. Vis. Pattern Recognit. (CVPR), pp. 580–587 (2014). https://doi.org/10.1109/CVPR.2014.81
23. Howard, A.G., Zhu, M., Chen, B., Kalenichenko, D., Wang, W., Weyand, T., et al.: MobileNets: Efficient convolutional neural networks for mobile vision applications. arXiv preprint arXiv:1704.04861 (2017). https://arxiv.org/abs/1704.04861
24. Liu, W., Anguelov, D., Erhan, D., Szegedy, C., Reed, S., Fu, C.Y., Berg, A.C.: SSD: Single shot multibox detector. In: Eur. Conf. Comput. Vis. (ECCV), pp. 21–37 (2016). https://doi.org/10.1007/978-3-319-46448-0_2
25. Simonyan, K., Zisserman, A.: Very deep convolutional networks for large-scale image recognition. In: Int. Conf. Learn. Represent. (ICLR) (2015). https://arxiv.org/abs/1409.1556
26. Zhao, Q., Sheng, T., Wang, Y., Tang, Z., Chen, Y., Cai, L., Ling, H.: M2Det: A single-shot object detector based on multi-level feature pyramid network. In: Proc. AAAI Conf. Artif. Intell. **33**(1), 9259–9266 (2019). https://doi.org/10.1609/aaai.v33i01.33019259
27. Chowdhury, A., Singh, A.K., Pal, A.: Efficient detection of agricultural diseases using EfficientDet. Comput. Electron. Agric. **188**, 106288 (2021)
28. Saponara, S., Elhanashi, A. Gagliardi,, A.: Real-time video fire/smoke detection based on CNN in antifire surveillance systems. J. Real-Time Image Process. **18**, 889–900 (2021). https://doi.org/10.1007/s11554-020-01044-0
29. Kim, Jin Su, Min-Gu Kim, and Sung Bum Pan. "A study on implementation of real-time intelligent video surveillance system based on embedded module." EURASIP Journal on Image and Video Processing 2021.1 (2021): 35
30. Tan, M., Le, Q.V.: EfficientNet: Rethinking model scaling for convolutional neural networks. In: Int. Conf. Mach. Learn. (ICML), pp. 6105–6114. PMLR (2019)
31. Tan, M., Pang, R., Le, Q.V.: EfficientDet: Scalable and efficient object detection. In: Proc. IEEE/CVF Conf. Comput. Vis. Pattern Recognit. (CVPR), pp. 10781–10790 (2020)

32. Wang, Z., Zhang, Y., Liu, H.: Real-time medical object detection with EfficientDet in edge AI systems. IEEE Access **9**, 123456–123467 (2021)
33. Zhou, Y., Li, Y., Fang, B.: Lightweight and real-time object detection for autonomous vehicles using EfficientDet. Sensors **21**(14), 4719 (2021)
34. Zhou, X., Wang, D., Krähenbühl, P.: Objects as points. arXiv preprint arXiv:1904.07850 (2019). https://arxiv.org/abs/1904.07850
35. Li, Y., Li, Y., Wang, Y., Zhang, J.: Efficient object detection using optimized deep learning frameworks: A review. Sensors **22**(4), 1503 (2022). https://doi.org/10.3390/s22041503
36. Liu, H., Wang, Y., Yang, M.: Recent progress in object detection: A review. J. Vis. Commun. Image Represent. **74**, 102827 (2020). https://doi.org/10.1016/j.jvcir.2020.102827
37. Ren, S., He, K., Girshick, R., Sun, J.: Object detection with region proposal networks. IEEE Trans. Pattern Anal. Mach. Intell. **39**(6), 1137–1149 (2017). https://doi.org/10.1109/TPAMI.2016.2577031
38. Zhang, X., Sun, H., Yang, X.: Advances in object detection for intelligent surveillance systems: A survey. Multimed. Tools Appl. **80**(10), 15227–15249 (2021). https://doi.org/10.1007/s11042-021-10547-8
39. Zou, Z., Shi, Z., Guo, Y., Ye, J.: Deep learning-based object detection: A comprehensive review. Artif. Intell. Rev. **56**, 1681–1710 (2023). https://doi.org/10.1007/s10462-022-10240-4
40. Chen, K., Pang, J., Wang, J., Xiong, Y., Li, X., Sun, J., Loy, C.C.: Hybrid task cascade for instance segmentation. In: Proc. IEEE/CVF Conf. Comput. Vis. Pattern Recognit. (CVPR), pp. 4974–4983 (2019)
41. Girshick, R.: Fast R-CNN. In: Proc. IEEE Int. Conf. Comput. Vis. (ICCV), pp. 1440–1448 (2015)
42. He, K., Gkioxari, G., Dollár, P., Girshick, R.: Mask R-CNN. In: Proc. IEEE Int. Conf. Comput. Vis. (ICCV), pp. 2961–2969 (2017)
43. Lin, T.Y., Dollár, P., Girshick, R., He, K., Hariharan, B., Belongie, S.: Feature pyramid networks for object detection. In: Proc. IEEE Conf. Comput. Vis. Pattern Recognit. (CVPR), pp. 2117–2125 (2017)
44. Liu, Y., Otálora, S., Atzori, M., Müller, H.: Fast and accurate medical image object detection with attention-guided region proposal networks. Comput. Med. Imaging Graph. **83**, 101748 (2020)
45. Ren, S., He, K., Girshick, R., Sun, J.: Faster R-CNN: Towards real-time object detection with region proposal networks. In: Adv. Neural Inf. Process. Syst. (NeurIPS), pp. 91–99 (2015)
46. Tian, Z., Shen, C., Chen, H., He, T.: FCOS: Fully convolutional one-stage object detection. In: Proc. IEEE/CVF Int. Conf. Comput. Vis. (ICCV), pp. 9627–9636 (2019)
47. Uijlings, J.R.R., van de Sande, K.E.A., Gevers, T., Smeulders, A.W.M.: Selective search for object recognition. Int. J. Comput. Vis. **104**(2), 154–171 (2013)
48. Zitnick, C.L., Dollár, P.: Edge boxes: Locating object proposals from edges. In: Eur. Conf. Comput. Vis. (ECCV), pp. 391–405 (2014)
49. Chen, L.C., Papandreou, G., Kokkinos, I., Murphy, K., Yuille, A.L.: DeepLab: Semantic image segmentation with deep convolutional nets, atrous convolution, and fully connected CRFs. IEEE Trans. Pattern Anal. Mach. Intell. **40**(4), 834–848 (2018). https://doi.org/10.1109/TPAMI.2017.2699184
50. Ghiasi, G., Lin, T.Y., Le, Q.V.: NAS-FPN: Learning scalable feature pyramid architecture for object detection. In: Proc. IEEE/CVF Conf. Comput. Vis. Pattern Recognit. (CVPR), pp. 7036–7045 (2019). https://doi.org/10.1109/CVPR.2019.00721
51. He, K., Gkioxari, G., Dollár, P., Girshick, R.: Mask R-CNN. In: Proc. IEEE Int. Conf. Comput. Vis. (ICCV), pp. 2961–2969 (2017). https://doi.org/10.1109/ICCV.2017.322
52. Liu, S., Qi, L., Qin, H., Shi, J., Jia, J.: Path aggregation network for instance segmentation. In: Proc. IEEE Conf. Comput. Vis. Pattern Recognit. (CVPR), pp. 8759–8768 (2018). https://doi.org/10.1109/CVPR.2018.00913
53. Banbury, C.R., Zhou, C., Fedorov, I., Matas, R., Thakker, U., Gope, D., Reddi, V.J.: Micronets: Neural network architectures for deploying TinyML applications on commodity microcontrollers. arXiv preprint arXiv:2007.10319 (2021)

54. Canziani, A., Paszke, A., Culurciello, E.: An analysis of deep neural network models for practical applications. arXiv preprint arXiv:1605.07678 (2016)
55. David, R., Duke, J., Jain, A., Reddi, V.J., Jeffries, N., Li, J., et al.: TensorFlow Lite Micro: Embedded machine learning on TinyML systems. arXiv preprint arXiv:2010.08678 (2020)
56. Ghosal, S., Blystone, D., Singh, A.K., Ganapathysubramanian, B., Singh, A., Sarkar, S.: Explainable deep machine vision framework for plant stress phenotyping. Proc. Natl. Acad. Sci. **118**(6) (2021)
57. Jacob, B., Kligys, S., Chen, B., Zhu, M., Tang, M., Howard, A., et al.: Quantization and training of neural networks for efficient integer-arithmetic-only inference. In: Proc. IEEE Conf. Comput. Vis. Pattern Recognit. (CVPR), pp. 2704–2713 (2018)
58. Lin, J., Chen, W.M., Lin, Y., Cohn, J., Gan, C., Han, S.: MCUNet: Tiny deep learning on IoT devices. arXiv preprint arXiv:2007.10319 (2020)

Chapter 4
Understanding Localization

Abstract This chapter addresses the core issue of localization in computer vision, which involves estimating the position of objects in an image correctly—not just classifying them. The chapter considers commonplace methods to get to localization, including bounding box regression for coarse localization, instance segmentation (masking) for pixel-level accuracy and landmark detection for the purpose of finding the specific location of object parts. The chapter discusses the role of confidence indices to filter the detections and critical quality ground truth data for training and evaluating computer vision models. The chapter further provides evaluation metrics that quantify localization quality, such as the notion of Intersection over Union (IoU), and then the probabilistic foundations for models to quantify prediction uncertainty. The chapter discusses primary loss functions, including both Smooth L1 and IoU based losses, that are iteratively optimized for bounding box estimation. The chapter covers a wide range of forces that underpin the critical idea of localization so that the chapter serves to provide the foundational concepts needed to understand and then consider metrics useful for quantifying improvement in the accuracy of object localization for computer vision in both modern and deep learning systems.

Localization in deep learning refers to the process of precisely identifying the position of an object within an image or video frame. Unlike classification, which only determines what an object is, localization goes a step further by pinpointing where the object is located, typically using bounding boxes or segmentation masks. This capability is fundamental in applications such as autonomous driving, medical imaging, and surveillance, where knowing an object's exact position is as crucial as recognizing its class. The book Deep Learning for Object Detection and Localization explores the underlying principles, architectures, and techniques that enable machines to perform these tasks with high accuracy. Understanding localization requires a grasp of both traditional computer vision methods and modern deep learning approaches, as the field has evolved significantly with the advent of convolutional neural networks (CNNs) and transformer-based models.

A. Elhanashi and S. Saponara, *Deep Learning for Object Detection and Localization*,
https://doi.org/10.1007/978-981-95-4846-0_4

One of the earliest methods for object localization involved sliding window techniques, where classifiers were applied across different regions of an image to detect objects. However, these approaches were computationally expensive and inefficient for real-time applications. The introduction of region-based CNNs, such as R-CNN, marked a significant leap forward by using selective search to propose potential regions of interest before applying a CNN for classification and bounding box regression. With all its ameliorations, R-CNN was still rather slow because of its multi-stage nature. This, in turn, gave impetus to the development of Fast R-CNN and Faster R-CNN, which maximized speed and accuracy by sharing computations and incorporating RPNs. These developments paved the way for many modern localization systems, emphasizing end-to-end trainable systems for top performance.

Single-shot detectors such as YOLO (You Only Look Once) and SSD (Single Shot Multibox Detector) transformed object localization by eliminating the separate region proposal stage. These networks, instead, predict bounding boxes and class probabilities directly from the image in one single forward pass, making them considerably faster yet nearly as accurate. For example, by splitting the entire image into an array of grids, YOLO could predict bounding boxes in every grid cell, allowing detections in real time, even on devices of lower capacity. In contrast, SSD makes use of multi-scale feature maps to locate objects of various sizes and thus improves localization accuracy at different scales. These approaches stress the trade-off between speed and accuracy, a constant theme running through the deep-learning model design for localization. One other important aspect of localization is anchor boxes used as reference shapes to predict bounding boxes.

Anchors allow the model to learn variations in object sizes and aspect ratios, further strengthening the robustness of all detections. However, correctly choosing the number and dimensions of anchor boxes remains a big problem; poor choices can cause missed detections or provide too much overhead in computations. New advancements, namely anchor-free methods, try to remove these anchors altogether and predict the centers and sizes of objects directly. In particular, CenterNet and FCOS (Fully Convolutional One-Stage Object Detection) yield good results, showing that precise localization is possible without anchor-based priors. This, in turn, represents ongoing innovations in localization in deep learning.

In instance segmentation, the exact shape of objects must be identified in an image. Mask R-CNN-like applications extend Faster R-CNN by adding an extra subnet for pixel-wise segmentation masks that mark the object with precise boundaries. The need to localize quite finely is an essential concern in diagnosis and treatment planning in instances such as medical imaging. Likewise, panoptic segmentation combines semantic and instance segmentation, delivering the complete scene's object and background understanding. This is witness enough that localization has moved out of coarse bounding boxes to pixel-level precision under the pressure of ever-complex real-world tasks.

Loss functions have always dictated proper learning of bounding box prediction by the model, which makes one think if it can be totally ignored in the historical account of localization. Smooth L1 loss for bounding box regression takes care of limiting the influence of outliers, while IoU loss provides direct optimization criteria concerning the overlap of predicted and ground-truth boxes. GIoU- and DIoU-loss methods were more recently proposed to overcome drawbacks of (traditional) IoU, such as dealing with non-overlapping boxes and considering center-point distances. These recent proposals indicate a true topical interest in designing loss functions concerning precise localization as the end goal so that such models will behave well under varying conditions.

Data augmentation is vital in improving localization robustness, ensuring that models learn under varying conditions during the training phase. Random cropping, rotation, and color jitters prevent overfitting and help with further generalization. Careful consideration must be undertaken, ensuring that augmentations do not change object positions or introduce completely unrealistic artifacts. Some sophisticated augmentation methods involve copy-paste augmentation, wherein objects in one image are pasted into another, and this has been found to help in improving localization performances, especially in small datasets with very few annotated examples.

The prevalence of transformer-based models, such as DETR, has paved a new way for localization using self-attention in global context building. Contrasting with CNNs that follow hierarchical processing steps, DETR signals a new perspective in object detection as a set prediction problem, freeing the system from unwanted hand-designed elements such as anchor boxes or non-maximum suppression. Though promising, the pain behind deploying such models in resource-constrained environments lies in the bearing of these models on very-large-scale training data and computational resources. However, these models appear very helpful in modeling long-range dependencies, making them relevant in complicated scenes where objects interact in dynamic ways.

Real-world scenarios of localization applications include, but are not limited to, autonomous cars detecting pedestrians and obstacles, inventory systems in retail, and customer behavior systems. Applications in healthcare include localization for the detection of tumors or anomalies in medical scans, whereas in agriculture, applications include monitoring crop health and pest attacks. Systems often consider various parameters such as lighting changes and occlusion problems, which differ for each application; class imbalance also affects how one approaches classifying instances. Hence, the book might cover these applications more thoroughly and show the intricacies with which localization models are modified for differing domain requirements.

A bright future lies in localization in deep learning, aiming at greater efficiency, interpreting ability, and generalization. Lightweight architectures, NAS, and self-supervised learning are new trends ensuring that fewer labels are necessary and that models become more adaptable. In addition, we are witnessing emerging explainable AI techniques that would allow the practitioner to understand and thereby trust localization predictions. The challenges to be tackled in the coming days regarding ethics, bias, and real-time deployment will, in fact, require multidisciplinary collaboration.

4.1 Types of Localization (Bounding Box Regression, Masking, Keypoint Detection)

Bounding Box Regression, Masking (Instance Segmentation), and Keypoint Detection. Each method offers different granularity and complexity for localizing objects and are usually combined in some state-of-the-art object detection framework. Below follows an elaborate discussion on these forms of localization.

4.1.1 Bounding Box Regression

Bounding box regression remains the most common and most basic way of localizing objects in deep learning; the model predicts a rectangle's coordinates supposed to tightly envelop an object of interest inside the image.

- Definition: Given an image, a model would give out four coordinate values describing a bounding box - usually the (x, y) coordinates of the top-left corner, and width and height of the box. Unlike segmentation boxes, this gives a very coarse spatial extent of the object [1, 3, 4].
- Deep Learning Approach:Faster-RCNN, RetinaNet, and SSD are state-of-the-art object detectors that used CNNs for bounding box regression, where the networks were supposed to learn mapping from image features to bounding box coordinates through the regression layers with losses including smooth L1 and IoU-based loss [2].
- Challenges and Improvements: Conventional bounding box regression involves predicting the center and size of the box, and this can be imprecise if there is a large variance between the anchor boxes and the true object location. More recent proposals choose a refined approach, such as Side-Aware Boundary Localization (SABL), which predicts each side of the bounding box individually using different branches of the network. For better accuracy, the resolution is designed in two steps: bucket prediction and fine position identification. Such a design improves localization precision significantly with pathbreaking detectors [5, 7].
- Applications: Bounding box regression enjoys wide use in real-time object detection, autonomous driving, surveillance, and all those applications where fast yet reasonably precise object localization is needed [6].

4.1.2 Masking (Instance Segmentation)

Masking goes beyond localization and predicts a pixel-wise mask for each detected object, allowing for a rather fine appearance while delineating the shape of an object [8].

- Concepts: A binary mask over pixels uniquely identifies the pixels belonging to the object, in contrast with rectangular bounding boxes. This becomes very crucial when one wants to infer the shape of the object, as in medical images and image editing.
- Deep Learning Models: Mask R-CNN is among the most influential frameworks, an extension to Faster R-CNN with the addition of another branch for predicting segmentation masks on each Region of Interest (RoI). The mask branch predicts pixel-level segmentation mask for each detected object to do instance segmentation.
- Advantages: The mask gives a fine representation of objects, especially irregular in shape, or when the objects overlap. This mask analysis offers more minute analysis to downstream tasks such as object tracking and scene understanding.
- Disadvantages: Mask prediction is way more computationally expensive than bounding box regression and needs detailed annotations during training. Thus, such models must strike a balance between accuracy and speed if they are to be useful practically.

4.1.3 Keypoint Detection

Every object detection system needs keypoint detection to identify crucial points on items which include facial features, body joints and object components which help in precise pose analysis and structural evaluation [9].

- The detection objective involves finding specific points in a specific pattern which are known as keypoints. When people estimate human body poses, their models look for joints such as elbows and knees and wrists.
- Machine learning algorithms convert keypoint detection into a heatmap regression task through which networks produce probability maps that identify each keypoint location. The models OpenPose and HRNet have shown exceptional performance levels in detecting numerous keypoints at the same time.
- Keypoint detection serves as a fundamental capability for human-computer interaction as well as augmented reality and robotics and behavior analysis applications. Spatial details obtained through keypoint detection surpass the information available from bounding boxes or masks.
- The combination of object detection and keypoint detection in certain frameworks facilitates both pose estimation and object localization processes through one operation. The resulting integration delivers improvements in both system resilience and operational speed.

4.2 Confidence Score

In the context of detecting an item, an important metric is the confidence score, which expresses how certain a model is about the presence of an object in one of the proposed bounding boxes. It is a probabilistic measure that usually runs between 0 (no confidence) and 1 (complete confidence) [10]. A higher score means that the model is pretty sure that an object exists within the predicted region, whereas a lower score means the model is unsure or considers it a false positive. Confidence scores are equally important for:

- Discarding weak detections (against a threshold, say 0.5).
- Ranking predictions (for example, Non-Maximum Suppression).
- Improving calibration (from confidence to accuracy). In this section, we discuss how confidence scores are obtained and the detection role of confidence, while also exploring their relation to classification, localization, and other pipeline components.

4.2.1 Confidence Score Formulation

The confidence score, thereby, is computed with the following two components:

1. Objectness Probability [11]—The chances that the bounding box contains some object (vs background).
2. Classification Probability—The distribution of probabilities of the object classes (assuming the existence of an object).

Equation 4.1: Raw Confidence Score
In the most simple form, confidence is the product of objectness probability, p_{obj}, and the largest class probability, $\max(p_{class})$ Eq. (4.1):

$$C_{raw} = p_{obj} \times \max(p_{class}) \tag{4.1}$$

where:

- p_{obj}—Probability that the box contains an object (output objectness branch).
- p_{class}—Softmax output over all possible classes.

It is with the above-mentioned formula that YOLO and other single-shot detectors calculate confidence.

Equation 4.2: Calibrated Confidence Score
Nowadays, calibration after-prediction via temperature scaling, Platt scaling, etc., is done so that the confidence outputs better reflect true probabilities. Hence, the calibrated score can be defined as Eq. (4.2):

$$C_{calibrated} = \sigma\big(T \cdot \log(C_{raw})\big) \tag{4.2}$$

where:

- σ : Sigmoid function.
- T : Temperature parameter (trained on a small validation set).

Calibration is most critical when it comes to safety-critical scenarios such as autonomous driving, where overconfident predictions can be quite dangerous.

Equation 4.3: Bayesian Confidence Estimation
Alternative confidence-as-uncertainty Bayesian modeling methods exist (e.g., Monte Carlo Dropout, Deep Ensembles). Ideally, the confidence score is approximated as Eq. (4.3):

$$C_{Bayesian} = \mathbb{E}[C_{raw}] - \text{Var}(C_{raw}) \tag{4.3}$$

where:

- $\mathbb{E}[C_{raw}]$: Expected confidence over stochastic forward passes.
- $\text{Var}(C_{raw})$: Variance, representing epistemic uncertainty.

This approach is useful when the model must account for its own limitations in unfamiliar scenarios.

4.2.2 Role of Confidence in Detection Pipelines

- Thresholding the Detections
  Most detectors will use, and apply a threshold (τ) to eliminate bad detections Eq. (4.4):

$$\text{Retained Detections} = \{(B_i, C_i) \mid C_i \geq \tau\} \tag{4.4}$$

  Typically, thresholding occurs between 0.3 and 0.7, based on the task.
- Non-Maximum Suppression (NMS)
  NMS uses confidence scores to remove duplicate detections. The system will keep the highest confidence box and remove any other boxes with lower confidence that overlap Eq. (4.5).

$$\text{Suppress } B_j \text{ if } IoU(B_i, B_j) \geq \theta \text{ and } C_j < C_i \tag{4.5}$$

where:

– θ : IoU threshold (e.g., 0.5).

Having a high confidence threshold will increase the precision of the object detector, but reduce recall. The Precision-Recall curve (PR curve) is used to visualize this trade-off.

4.2.3 Improving Confidence Estimation

1. Label Smoothing
   Hard labels (i.e., 0 or 1) can yield too confident predictions. The method of label smoothing replaces hard/0/1 labels with soft labels (e.g., 0.9 instead of 1) to encourage improved generalization.
2. Test-Time Augmentation (TTA)
   Each instance is run through a number of augmented versions of itself, then the confidences across the augmented versions are averaged together. Thus, we can reduce the variance in predictions and improve trustworthiness.
3. Uncertainty-Aware Training
   E.g., Evidential Deep Learning (EDL) considers model confidence as a Dirichlet distribution so that estimates of uncertainty can also be more accurate.

4.2.4 Practical Considerations

1. Dataset Bias: While models may assign appropriate confidence to common classes, they may reserve uncertainty for rare classes.
2. Adversarial Attacks: Small perturbations of an object can artificially boost confidence scores.
3. Real-Time Constraints: The computation of confidence scores should not be slower than inference speed.

The confidence score is an essential part of object detection systems because it influences filtering, ranking, and uncertainty modeling. By understanding its formulation (Eq. 1-3) and calibration techniques, practitioners can build a more reliable and interpretable detector.

4.3 Probability

Probability theory forms the mathematical basis for uncertainty quantification. Probability concepts are at the very core of deep learning models involved in object detection and localization. To simplify the discussion, we will only consider probabilistic reasoning when working with real data and informed predictions, estimation of a model's confidence level, and the ambiguity found in visual recognition tasks should the model use probabilities. Figure 4.1 illustrates an example of a classification probability distribution.

- The Role of Probability in Object Detection
  In object detection, models can create both a detection of the presence of an object and the localization of the object via a bounding box. In each case, the model prediction is assigned a confidence score or probability estimation. For example, a neural network could give an output of:
  - Class probabilities (i.e., $P(\text{class} = \text{"car"}) = 0.92$)
  - Localization uncertainties (variance in the predicted bounding box coordinates)

  These probabilities give the models a mechanism to filter out the low confidence predictions via thresholding (e.g., Non-Maximum Suppression) and improve the reliability of the detection.

4.3.1 Key Probability Concepts

1. Conditional Probability
   In common object detection scenarios, models use probabilities giving us an indication of whether an object is present given some observed pixel data in the

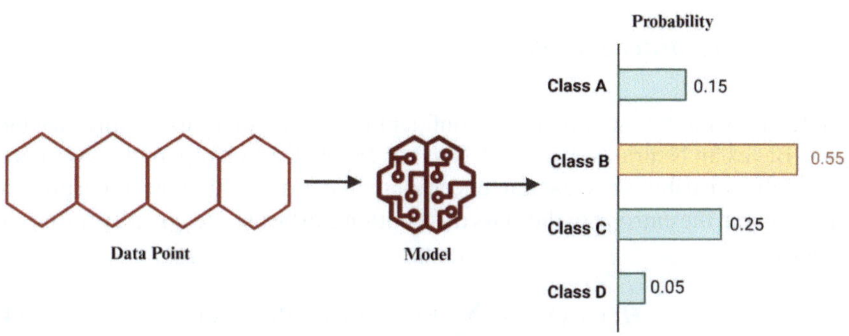

Fig. 4.1 An example of a classification probability distribution. For a given data point, the model assigns a probability score to each possible class (e.g., Class A, B, C, D), indicating its confidence. The predicted class is typically the one with the highest probability (in this case, Class B with 0.55)

image. Let x be the image patch and let y be the object class, we get conditional probabilities like:

a. The above is a consequence of Bayes' Theorem, Eq. (4.6):

$$P(y \mid x) = \frac{P(x \mid y) \cdot P(y)}{P(x)} \tag{4.6}$$

- $P(x \mid y)$ is the likelihood of observing the patch x if the object is y,
- $P(y)$ is the prior probability of the object class,
- $P(x)$ is the evidence (marginal probability of the patch).

Deep learning models fit $P(y \mid x)$ directly using discriminative classifiers such as convolutional neural networks (CNNs).

2. Probability Distributions for Localization
   Bounding box predictions are often represented as probability distributions over coordinates. Thus, for example, the predicted center.
   Deep learning models approximate $P(y \mid x)$ directly using discriminative classifiers like CNNs.
3. Probabilistic Distributions for Localization
   Bounding box predictions are often interpreted as a probability distribution over coordinates. For example, the predicted centre (c_x, c_y) of a bounding box may follow a Gaussian distribution Eq. (4.7):

$$p(c_x, c_y) = \mathcal{N}(\mu, \Sigma) \tag{4.7}$$

where μ is the mean (or expected location) and Σ is the covariance matrix which conveys the uncertainty. Modern detectors such as Gaussian YOLOv3 will explicitly model the distributions to offer more accurate localization.

4.3.2 Uncertainty Estimation

Deep learning models are often over confident in false predictions. To alleviate these issues, Bayesian Neural Networks (BNNs) or Monte Carlo Dropout treat weights as a probability distribution to capture uncertainty in metric predictions. For a predicted bounding box, the entropy of the class distribution can be used to quantify uncertainty Eq. (4.8) :

$$H(y \mid x) = -\sum_i P(y_i \mid x) \log P(y_i \mid x) \tag{4.8}$$

Probability theory has been the backbone for object detection as it allows models to quantify prediction level of confidence, model uncertainty in localisation, to decide to ignore uncertain detections. Following this, by functionalising probabilistic reasoning, deep learning can account for many sources of noise and inherent variation in real data which are viewed as nuisances in most engineering problems. In future work, it will be useful to apply more sophisticated probabilistic methods such as normalising flows or Bayesian neural networks to estimate uncertainty in a best-case scenario, and consequently increase robustness in autonomous vision systems.

4.4 Ground Truth

Ground truth in object localization and detection refers to the definitive annotations that describe the exact identified positions, boundaries, and classes of detected objects in images. Most importantly, we use ground truth as the baseline for training algorithms and the gold standard for testing and measuring model performance. In practice, ground truth is created by expert annotators that identify the exact bounds of the identified object with bounding boxes, segmentation masks, or other location-based indicators that quantify where the object of interest exists in each image [12]. Figure 4.2 shows the comparison between a model's detection result (brown line) and the ground truth bounding box (blue line). Depending on the quality of the ground truth data, the performance of detection models can vary significantly. It takes only a few poorly formed labels, overlapping boundaries, ambiguous boundaries, etc., for confusion to occur during the training process, and the model to learn incorrect representations. To assure reliability, all instances of an object in the dataset are labeled precisely with coordinates that describe the area the object exists in. Generally, bounding box annotations have top-left corner coordinates (x_1, y_1) and bottom-right corner coordinates (x_2, y_2) for rectangular bounding boxes. These values describe the spatial extent of the object while also being necessary for evaluation metrics, such

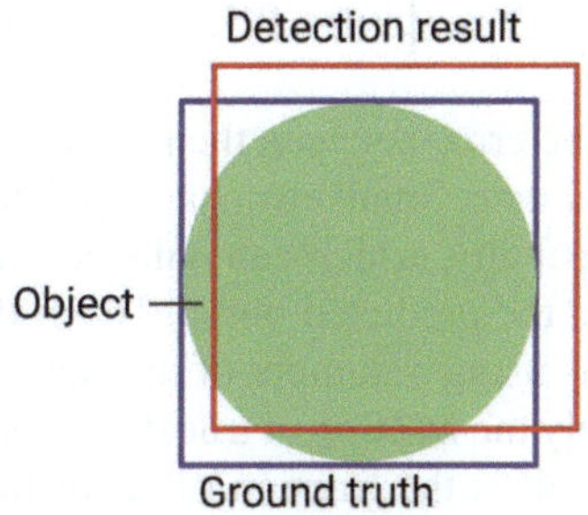

Fig. 4.2 comparison between a model's detection result (brown line) and the ground truth bounding box (blue line). The ground truth represents the perfect, human-annotated location of the object, against which the detection result is evaluated

as IoU. One of the important equations for comparing predicted bounding boxes to the ground truth, is called IoU. The IoU is defined as Eq. (4.9):

$$IoU = \frac{\text{Area of Overlap}}{\text{Area of Union}} \tag{4.9}$$

In this context, the "Area of Overlap" refers to the portion of the predicted and actual bounding boxes that overlap, and the "Area of Union" refers to the combined area occupied by both bounding boxes. Therefore, an IoU score that is closer to 1 represents a more similar prediction and ground truth; thresholds (e.g. 0.5 or 0.7) can be used to determine whether or not the prediction can be considered a detection. The IoU score is used not only in training (where it may affect the loss calculation) but also in evaluation (for example, mean Average Precision (mAP) depends directly on IoU scores).

When annotating ground truth data, annotators must take into consideration the object scales, occlusions and ambiguities associated with the edge of objects, etc. It is important that annotators are consistent, regardless of the guidelines used, potentially affecting the model learning process in an unintended subjective manner. Furthermore, ground truth typically includes class labels which allow the system to learn where detected objects are located, but also which class the detected object fits into. The coupled nature of location and classification adds more complexity, requiring a combination of strong annotation tools and thoroughly verified processes.

Accuracy in ground truth directly impacts loss functions during model training. One common approach combines classification loss with localization loss—a method that penalizes the model for both misclassifications and poor bounding box predictions. The localization loss is sometimes computed using a smooth L1 loss function, which is less sensitive to outliers than a traditional L2 loss. The smooth L1 loss for a bounding box coordinate can be formulated as Eq. (4.10):

$$L_{reg}(x) = \begin{cases} 0.5x^2, & \text{if } |x| < 1 \\ |x| - 0.5, & \text{otherwise} \end{cases} \tag{4.10}$$

In this case, x refers to the error between the predicted coordinate and the corresponding ground truth coordinate. Small errors are applied a quadratic penalty, large errors are applied a linear penalty, which can balance sensitivity and robustness in training. During evaluation, the predicted outputs from the model are compared to the ground truth annotations using a number of metrics, two of which are precision and recall. Precision is simply the fraction of true detections over all of the detections made by the model and recall is the fraction of all of the actual objects that were detected. Both metrics are defined Eqs. (4.11) and (4.12):

$$\text{Precision} = \frac{\text{True Positives}}{\text{True Positives} + \text{False Positives}} \tag{4.11}$$

$$\text{Recall} = \frac{\text{True Positives}}{\text{True Positives} + \text{False Negatives}} \tag{4.12}$$

A high precision value tells us that if the model predicts an object it is likely correct, while a high recall value indicates that if there are objects there, the model does a good job detecting most of them. The relation between precision and recall values is often summarized in the F1 score which is the harmonic mean of precision and recall. The F1 score is helpful when taking a balanced view of model performance.

It cannot be overstated how important ground truth data is. It is the initial building block to examine quantitative model performance but it also takes on an important role when you are modifying the architecture of your model or training paradigms. In terms of object detection systems for research or applied purposes the reliability of the object detection or localization system is a function of the quality or rigor in which ground truth has been defined, annotated and maintained. If high quality ground truth data is maintained it will support more objective evaluation of models and reliable performance once models are deployed to real world applications/contexts and facilitate to some degree some extension spatially of the field.

4.4.1 Structure of Ground Truth Data

The ground truth data is usually comprised of:

- Class labels: to indicate the class (e.g., person, car, dog).
- Bounding box coordinates
- Optional metadata: confidence score (if estimated by hand), segmentation masks, or occlusion information.

There can be many objects in a single image, hence the ground truth is often represented as a list of tuples per image.

Bounding Box Formats
Bounding boxes are important for localization, there are two primary formats for bounding boxes Eqs. (4.13) and (4.14):

1. Corner Coordinates Format (used in PASCAL VOC, COCO):

$$(x_{min}, y_{min}, x_{max}, y_{max}) \tag{4.13}$$

- x_{min}, y_{min}: are the top-left coordinate of the box
- x_{max}, y_{max}: are the bottom-right coordinate of the box

This defines a box from the top-left to the bottom-right pixel.

2. Center Coordinates Format

$$(x_{center}, y_{center}, w, h) \tag{4.14}$$

- x_{center}, y_{center} : are the coordinates of the center of the box
- w : width of the box
- h : height of the box

4.4.2 Ground Truth Matching

During training, the predictions are compared to the ground truth in order to calculate losses. Predictions are matched with ground truth boxes using Intersection over Union (IoU) Eq. (4.15):

$$IoU(A, B) = \frac{|A \cap B|}{|A \cup B|} \tag{4.15}$$

where A and B are the predicted and ground truth bounding boxes. The predicted and ground truth boxes are compared at a threshold (e.g. 0.5) to define whether a prediction is a true positive.

Loss Functions and Ground Truth
There are two losses that have ground truth during training:

- Classification loss—compares predicted class scores to the ground truth label (e.g. cross-entropy).
- Localization loss—compares predicted bounding box to ground truth (e.g. Smooth L1 or IoU loss).

Smooth L1 loss for localization Eq. (4.16):

$$L_{loc}(t, t^*) = \sum_{i \in \{x,y,w,h\}} smooth_{L_1}(t_i - t_i^*) \tag{4.16}$$

where:

- t are the predicted bounding box offsets
- t^* are the ground truth offsets

The smooth function is defined as Eq. (4.17):

$$smooth_{L_1}(x) = \begin{cases} 0.5x^2, & \text{if } |x| < 1 \\ |x| - 0.5, & \text{otherwise} \end{cases} \tag{4.17}$$

4.4.3 Challenges with Ground Truth

- Annotation errors: Annotation mistakes: Human mistakes while labeling ground truth can mislead the model.
- Ambiguity: If there is overlap due to other object or occlusions, it becomes unclear what is being annotated.
- Class imbalance: Some classes may be too small in ground truth and would require some kind of weighted loss or data augmentation method.

4.4.4 Tools for Ground Truth Annotation

A number of different tools are used to create ground truth:

- LabelImg: Manual bounding box annotation.
- CVAT: Supports segmentation and tracking.
- Roboflow, VGG Image Annotator: Roboflow, VGG Image

Annotator (supports collaboration and annotations). Export annotations to formats such as Pascal VOC (XML), COCO (JSON), and YOLO (TXT). Each of these formats need specific preprocessing to be used with a deep learning model.

4.5 Model Selection

Choosing the appropriate deep learning architecture for object detection and localization can be challenging because of the trade-offs between accuracy, speed, and computational load. The perfect system depends on the application because different applications have different metrics of performance that can be prioritized. For example, high-fidelity detections like in medical imaging or satellite imaging may favor architectures with higher accuracy while allowing slower speeds of inference applied in multi-stages; the deep learning model proposes some regions and subsequently refines the regions enough to allow accurate localization even for high-precision detections with complex or overlapping objects, given embroiled with higher computational load and rendered lesser suitable for real-time. On the other hand, as with autonomous robotics or a live streaming video feed as the input source, requiring real-time processing to interpret frames would require lightweight architectures that inferred a location and classification of an object, in this case, the inference over the input happens with one pass that must consider speed and accuracy trade-offs with limited computational resources. Further advances have optimized these architectures by promoting the elimination of predefined parameters and enabling training with relative less work and good performance [13].

Another very prominent consideration in choosing a model will be the dataset [14]. You can typically attempt a deeper, more complex architecture when there is an ample collection of data that is diverse, as that data can help capture variances in the object's appearance. For smaller datasets or when working under stricter resource constraints, more simplistic models may be necessary since any complicated architecture may be at risk of overfitting and even resource over usage. If different architectures didn't already create use cases, the introduction of more attention-inspired mechanisms (or other attention mechanisms) has given way to possibilities for relating to many well-cluttered scenes [15]—where understanding the global context often allows for better reliability of detection. The final decision of the model is predicated on a solid understanding of the task requirements, performance specifications, and hardware constraints. By judiciously considering these issues, practitioners will be able to select an architecture that is right and only the architecture that is right for your use case.

References

1. eInfochips: Understanding object localization with deep learning. Available online: https://www.einfochips.com/blog/understanding-object-localization-with-deep-learning/ (accessed on 26 September 2025)
2. Rosebrock, A.: Object detection bounding box regression with Keras, TensorFlow, and deep learning. PyImageSearch (2020). Available online: https://pyimagesearch.com/2020/10/05/object-detection-bounding-box-regression-with-keras-tensorflow-and-deep-learning/ (accessed on 26 September 2025)
3. Fiveable: Bounding box regression study guide. Available online: https://library.fiveable.me/images-as-data/unit-9/bounding-box-regression/study-guide/fOsrufrtx1skSwHh (accessed on 26 September 2025)
4. Zhang, A., Lipton, Z.C., Li, M., Smola, A.J.: Bounding boxes. In: Dive into Deep Learning. Available online: https://d2l.ai/chapter_computer-vision/bounding-box.html (accessed on 26 September 2025)
5. Zhu, C., He, Y., Savvides, M.: Side-aware boundary localization for more precise object detection. Available online: https://deeplearn.org/arxiv/106972/side-aware-boundary-localization-for-more-precise-object-detection (accessed on 26 September 2025)
6. Girshick, R., et al.: Paper at NeurIPS 2013. Available online: https://proceedings.neurips.cc/paper_files/paper/2013/file/f7cade80b7cc92b991cf4d2806d6bd78-Paper.pdf (accessed on 26 September 2025)
7. Zhu, C., He, Y., Savvides, M.: Side-aware boundary localization for more precise object detection. Available online: https://deepai.org/publication/side-aware-boundary-localization-for-more-precise-object-detection (accessed on 26 September 2025)
8. Carvalho, Osmar Luiz Ferreira de, et al. "Instance segmentation for large, multi-channel remote sensing imagery using mask-RCNN and a mosaicking approach." Remote Sensing 13.1 (2020): 39
9. Petrakis, Georgios, and Panagiotis Partsinevelos. "Keypoint detection and description through deep learning in unstructured environments." Robotics 12.5 (2023): 137
10. Wan, Sheng, et al. "Confnet: predict with confidence." 2018 IEEE International Conference on Acoustics, Speech and Signal Processing (ICASSP). IEEE, 2018
11. Feng, Di, et al. "A review and comparative study on probabilistic object detection in autonomous driving." IEEE Transactions on Intelligent Transportation Systems 23.8 (2021): 9961–9980

12. Foody, G.M. Ground Truth in Classification Accuracy Assessment: Myth and Reality. Geomatics 2024, 4, 81–90. https://doi.org/10.3390/geomatics4010005
13. Ding, Jie, Vahid Tarokh, and Yuhong Yang. "Model selection techniques: An overview." IEEE Signal Processing Magazine 35.6 (2018): 16–34
14. Chen, Xue-Wen, and Xiaotong Lin. "Big data deep learning: challenges and perspectives." IEEE access 2 (2014): 514–525
15. Shrestha, Rashmee, et al. "Object Recognition in a Cluttered Scene." Data Engineering for Smart Systems: Proceedings of SSIC 2021. Singapore: Springer Singapore, 2021. 71–76

Chapter 5
Applications of Object Detection and Localization

Abstract The chapter addresses changes brought about due to object detection and localization with deep learning. Models like YOLO, Faster R-CNN, and SSD allow machines to find and classify objects in images and videos accurately. The discussion is divided into three primary applications of these systems. In autonomous driving, these systems are foregrounded because they allow machines to identify pedestrians, vehicles, and traffic signs, which is needed to guide vehicles through real-time navigation safely. In surveillance systems, object detection and localization systems allow for automated threat detection, traffic observation, and retail observation, which ultimately enhance safety in private and public spaces. In medical imaging, object detection and localization can help in diagnostic activities by identifying tumors, segmenting organs, and reviewing scans for computer aided diagnosis, thereby helping to lessen the time and improve the accuracy of diagnoses. While there are some challenges such as limitations on computational abilities and reservations about informing society about data privacy, work continues in these areas to stretch these boundaries of intelligent systems and advance towards a safer, more efficient, and automated engagement with our physical world.

Deep learning for object detection has changed the way machines can view and interpret our visual stimulion. By combining convolutional neural networks (CNNs) with more advanced learning techniques, modern object detection systems can now detect and localize numerous objects in an image or video frame with significant accuracy. Machines can now go from image classification to understanding the spatial coordinates of multiple objects in the past, the historical context in how objects can affect each other, and many other relevant dimensions of the scene. There has been a huge shift and movement from traditional image processing and to deep learning type approaches that make the task of detecting objects faster and much more accurate (including in complicated and changing landscapes in a video). The forward progress made with deep learning models is seen in perplexing improvements to model based object detection. Object detection models, like YOLO (You Only Look Once), Faster R-CNN, and SSD (Single Shot MultiBox Detector) [1] have become

A. Elhanashi and S. Saponara, *Deep Learning for Object Detection and Localization*,
https://doi.org/10.1007/978-981-95-4846-0_5

staple models that form the fundamentals of object detection. These models have have leaked the requirement for building kets of image feature extraction manually (or encoding them) and are also able to learn directly from shared data, which means that they can be used in more potential contexts, without a limited step of shared information before seeing the examples. Many models are capable of accommodating occlusion, various dimensions of objects, and real-time detection. The combination of ample annotated data and GPU-enabled training environments have helped speed the process. Therefore, object detection with deep learning continues to advance the limits of computer vision, and pave the way for intelligent systems interacting with the physical world.

5.1 Autonomous Driving

Autonomous driving is one of the most disruptive technologies driving the future of transportation [2]. The successful deployment of autonomous vehicles (AVs) goes beyond purely climatic concerns and is tied to their ability to accurately and instantaneously perceive and understand the environment around them. Object detection, simply put, is the task of localizing and classifying objects in image or video streams relevant to that perception. With the emergence (and continuing advancements) of deep-learning models, most prominently convolutional neural networks (CNNs) and various iterations of this architecture, object detection is no longer bound by the older methods of traditional and heuristic-based computer vision, and can instead handle many classes of objects with high fidelity [3]. As we look to develop reliable autonomous driving solutions, this is a significant advancement.

In an autonomous driving system, object detection is the most foundational layer for all higher level tasks, including path planning, avoidance, and decision making. The classes of objects that need to be detected are extensive; they span pedestrians, vehicles, traffic signs, cyclists, road/route boundaries, and obstacles. These deep learning models are trained on massive datasets to recognize the numerous object classes in a variety of conditions—day vs night, rain vs fog, urban vs rural—and would pose significant challenges to any deployment of autonomous vehicles in the real-world.

There are typically two types of learning based object detection models; two-stage detectors, and one-stage detectors. Two-stage detectors first surgically determine where an object may be present, and then classify that location, with models in the Region-based CNN (R-CNN) family being a common example. While it is likely to return a higher level of accuracy, the method is computationally expensive and therefore may be impractical for real-time applications. On the other end of the spectrum, one-stage detectors such as YOLO (You Only Look Once) or SSD (Single Shot MultiBox Detector) perform the modes of detection and classification simultaneously. This can greatly lessen local processing time, and is particularly important for real-time applications such as autonomous driving. The new set of models, referred to as the "new hybrid model", have been able to embrace speed and accuracy, making them applicable on embedded systems for autonomous vehicles.

One of the most relevant types of object detection in the context of autonomous driving applications is pedestrian detection [4]. The pursuit of pedestrian safety is important, thus autonomous vehicles need to be able to selectively trigger quick and accurate detection of pedestrians to prevent accidents from occurring. Deep learning models are likely able to deliver a model which can discriminate changes in labels of pedestrians, even in crowded backdrop scenarios or even in foreshortened views, e.g. when the pedestrian is in a parked car or stopped pedestrian. The task is significantly easier when provided multiple labelled datasets on pedestrian data such as in Caltech Pedestrian Dataset.

Another important area of application is the detection of vehicles. Autonomous vehicles need to be able to detect surrounding vehicles, including cars, trucks, buses, motorcycles, and bicycles, in order to navigate without incident. The deep learning model for vehicle detection is often used in conjunction with a tracking algorithm to make sense of the other vehicle's trajectory over time. It is important for the autonomous vehicle to understand the dynamics of the other vehicle such that it anticipates planned actions like changing lanes or suddenly stopping. The detection algorithm can also help detect emergency vehicles which may require special consideration like having to yield or make way.

Detecting and recognizing traffic signs is another important application of object detection in relation to autonomous driving [5]. Traffic signs inform and guide drivers with rules and operating information like speed limits, stop signs, pedestrian crossings, and road work signage. Artificial neural networks [6] allow autonomous vehicles (AV) to detect traffic signs from distance and motion, while classifying those signs, including recognizing certain traffic signs even under suboptimal lighting and weather conditions. This improves the ability of autonomous vehicles to adhere to relevant traffic laws and act in compliance accordingly. There are systems where the object detection capabilities of artificial neural networks are combined with OCR (optical character recognition) for reading variable message signs and electronic displays that are on the road.

Detection of cyclists is also important since cyclists are vulnerable road users and therefore its important to get their safety right. Detecting cyclists comes with different challenges as they are smaller, faster, and change appearance. Nevertheless, since deep learning models can rely on a specific dataset, they will be able to distinguish bicycles away from other possible objects and track their individual motion trajectories. If we accurately detect cyclists, autonomous vehicles will be able to give them the appropriate distance away from safety and predict their erratic movements. This can drastically lessen the chances of a collision.

In addition to the other primary object detection element, there are also many more objects detection can assist with, such as road boundaries [7], lane markings, obstacles on the road, or hazards, such as debris, construction cones, or animals. While semantic segmentation [8] usually delineates the entire road scene, object detection only shows bounding boxes with the specific object that we are concerned with. Instead of simply seeing the object in the road scene, we acknowledge it as a particular detected and documented object. The holy grail of enhancement is when detection results are fused using sensors such as lidar [9], radar, ultrasonic etc., allowing the vehicle to create an understanding of what is taking place outside the car at any given time, while allowing for a possibility of an autonomous vehicle working in a complicated driving scenario of image compression and activity to still make safe decisions about how to proceed.

Real-time processing is a central issue and obstacle in deep learning object detection applications, particularly within autonomous driving. Models must adhere to tight latency requirements to take timely actions after detection. A visible trade-off exists between accuracy and computational costs, but model pruning, quantization methods, and other lightweight model structures can help mitigate this trade-off. Models can also be executed in real-time through a dedicated hardware accelerator on the vehicle (GPUs, TPUs, etc.) to allow enough inference time for the autonomous driving system. Underlying robustness of object detection systems offers another area of research once incorporated into autonomous driving fleets. Models need to perform too many different environments and lighting conditions, and adverse operating conditions through occlusions, shadows, noise, etc. Many approaches to improving robustness of models include using data augmentation, domain adaptation, and ensembling. In other contexts, inferring uncertainty in the detection output can be helpful for the vehicle's control system to help understand the level of reliability of the detection and act more conservatively when necessary. Another high impact application of object detection in autonomous driving is advanced driver-assistance systems (ADAS). ADAS systems leverage a very heavy reliance on real-time detection of objects around the vehicle to execute functions like automatic emergency brake, For example, when a vehicle ahead brakes suddenly, the temporary obstruction will call for automatic braking to avoid a collision. When vehicles or pedestrians are detected in the blind spots, the driver can be alerted to enhance their situational awareness and reduce the likelihood of an accident.

In addition to safety critical applications, object detection allows for comfort and convenience applications. Object detection will enable autonomous parking systems to find open parking spots and detect obstacles while positioning the vehicle into the parking spot. Object detection will complement the detection of traffic lights and road signals, allowing a vehicle to navigate intersections in a more seamless manner. The combination of object detection and the use of semantic mapping will allow vehicles to understand complex urban areas which includes sidewalks, crosswalks, and bus stops.

In addition to the safety critical applications, AI methods such as reinforcement learning [10] and sensor fusion will also serve to augment object detection approaches, for instance detected objects can be incorporated into predictive models to determine how the other road users are likely to behave. This form of anticipatory behaviour, coupled with the ability to assess the other's speed and distance from the vehicle, can allow for the vehicle to make proactive and not just reactive plans and contribute to safety and efficiency overall. In the future, the ongoing refinement of deep learning models for object detection in autonomous driving will carry the industry one step closer to fully autonomous vehicles. We will see trends

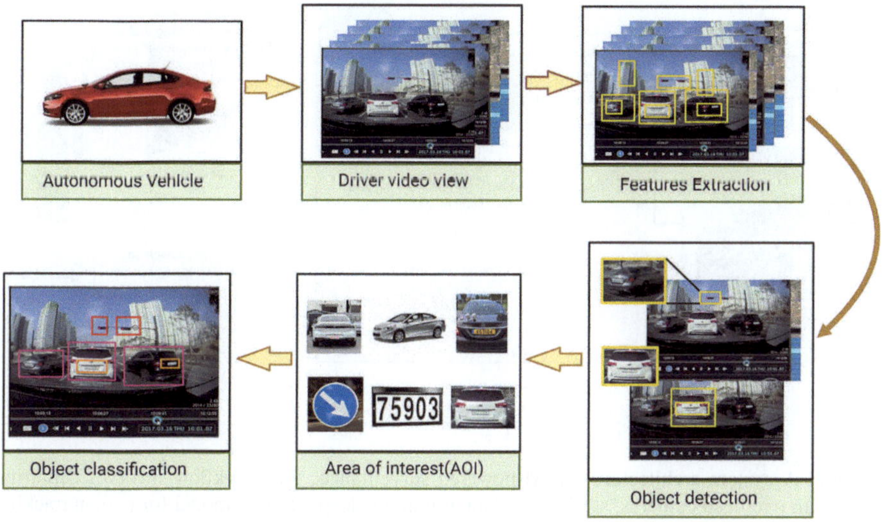

Fig. 5.1 Key computer vision tasks for autonomous vehicles. The system processes the driver's video view to perform feature extraction, object detection, classification, and define areas of interest (AOI) for safe navigation

with continued usage of 3D object detection using lidar and radar point clouds, multimodal sensor fusion [11], and transformer models that have shown potential in learning long-range dependencies, making better use of newly available data sources, and improving detection ability. Ongoing work is also being done to develop and evaluate end-to-end systems that can optimize perception, prediction and planning together under a single learning framework. This will create a more complete and efficient autonomous system. Figure 5.1 illustrates the key computer vision tasks for autonomous vehicles.

5.2 Surveillance Systems

Surveillance systems have changed significantly in the last decade due to advances in computer vision and artificial intelligence. The biggest game-changers in this area are deep learning and object detection methods, which allow for automated and real-time video assessment, leading to smarter, faster, and cheaper surveillance. In particular, deep learning—especially convolutional neural networks (CNNs)—have greatly changed the way that objects, people, and activities are identified, classified, and tracked in surveillance videos.

Object detection is the foundation of intelligent surveillance, since it involves identifying the presence of object instances (people, cars, suspicious items, etc.) in images or video. Traditionally, surveillance required human labor and a lot of

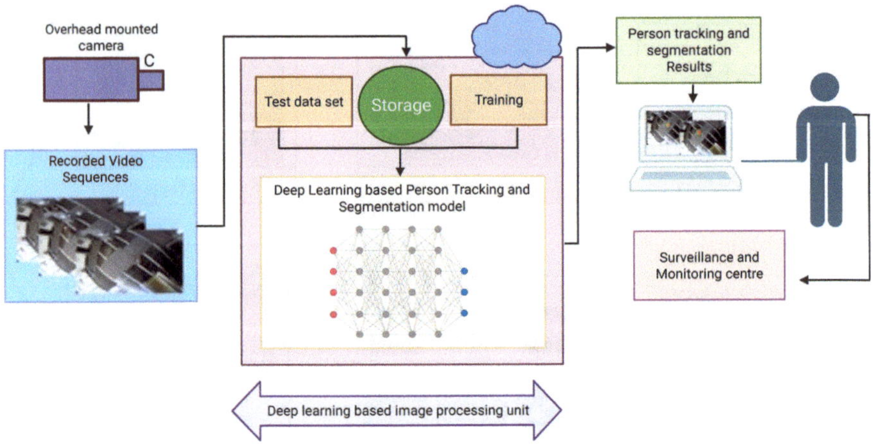

Fig. 5.2 A system architecture for AI-powered surveillance. The workflow involves capturing video with overhead cameras, storing the data, training a deep learning model for person tracking and segmentation, and processing the results in a monitoring centre

human error to extend the monitoring time, and by extension it was labor intensive to humans to process the vast amounts of subset video continuously. When deep learning based methods are applied to surveillance, it allows for processed data to create object instances and subsequently trigger alerts automatically. The applicability of this type of system extends beyond security, safety and/or law enforcement applications to traffic assessment, retail analysis and beyond. Figure 5.2 shows the system architecture for AI-powered surveillance applications.

5.2.1 Key Applications of Deep Learning and Object Detection in Surveillance Systems

1. Public Safety and Crime Prevention: Deep learning provides capabilities for detecting unusual behaviors or suspicious objects (e.g., an unattended bag), as well as monitoring potential crowd gatherings in real-time. Police and security forces have been deploying these capabilities in public areas such as airports, sports venues, and subway stations, to better understand the activities taking place around them and prevent potential crimes before they occur.
2. Traffic Monitoring and Management: Intelligent traffic surveilance and monitoring are one of the applications of object detection to surveil vehicles, detect vehicle offenses (i.e., speeding, illegal turning), and accidents. Some applications are able to count vehicles, classify the type of vehicles (cars, trucks, bicycles), and regulate the timers on traffic lights to reduce vehicle congestion and promote safer road conditions.

3. Retail Analytics and Loss Prevention: Retailers utilize center-mounted (interior) surveillance paired with deep learning technology to assess customer behavior, track foot traffic, and identify shoplifting behavior in real-time. Object detection allows the surveillance systems to differentiate between consumers and employees in the store, monitor hand movements of people to identify potentially suspicious movements, and automatically alert security personnel [12] .
4. Industrial Safety Monitoring: In highly controlled industrial environments, surveillance systems can detect the proper use of personal protective equipment by employees (such as helmets and safety vests), unsafe techniques used by employees, or people entering restricted areas, thus reducing the number of workplace incidents by intervening before accidents occur.
5. Security for Smart Homes and Buildings:Surveillance for consumer and business purposes uses object detection to recognize visitors, detect intrusions, and distinguish between people, pets, or objects. Integration with smart home systems can automate alerts, including response actions such as locking doors or switching on lights.
6. Healthcare and Care for the Elderly: Surveillance in hospitals and care homes uses deep learning to track patient movement, recognize falls, and confirm compliance with Hygiene protocols. Object detection can send alerts to staff if a patient requires help quickly, improving patient care.
7. Event and Crowd Control Security: Crowd control for large public events deploy the benefits of intelligent surveillance to monitor and measure crowd density, identify aggressive behavior, and confirm emergency route exits are clear. Continuous analysis in real-time can help event organizers manage the crowd and respond quickly [13].
8. Toll and Parking Automation: Object detection can recognize vehicles, read license plates, and automate toll collection or parking payments. Reduced congestion at toll booths, reduced physical contact and fraud detection at parking lots.

5.2.2 Challenges and Limitations

While deep learning and object detection have significant potential for surveillance systems, there remain challenges and constraints associated with working with IoT devices and cloud-based applications.

- Computational and Energy Resources for IoT Devices: Many surveillance cameras and sensors are deployed as limited processing power and energy sources as IoT edge devices. Even though deep learning models have various classes of lightweight algorithms that can run locally, the complex nature of real-time deep learning inference may still require substantial processing. Lightweight algorithms come with tradeoffs, and the accuracy and capabilities of the models can be limited, especially if certain subtle or rare events are being observed [14].

- Bandwidth and Latency to Cloud-Based Servers: Streaming high resolution and continuous video data from multiple cameras to cloud servers has demands for considerable bandwidth. In many settings where surveillance systems are deployed, the network capacity is generally constrained, especially in remote areas or part of a dense urban environment. We will often see limitations where latency compromises timely alerts and response—particularly immediate action in time-critical settings [15].
- Informed Consent and Data Security Concerns: Surveillance systems component solutions have access to, or collect personal identifiable information in a public or semi-public setting. As surveillance system data flows from field sources to servers, there are considerable risks associated with cloud processing and storage of what are often deemed sensitive data, and any form of data breach can cause significant harm in terms of risk only—any monitoring or capture of private data, especially in surveillance systems, will motivate legitimate concerns of data risks of further breeches. European regulations like the GDPR impose obligations of compliance and legal liability for data breach; adding to the complexity of deploying and managing data with surveillance systems.
- Model Generalization and Bias: Deep learning models may generalize poorly to new or unseen data if the model has been trained with a limited or biased dataset. Surveillance environments may vary significantly in terms of lighting, on-location weather, and/or the angle of camera capture. A model trained on a specific environment may generate vast number of alarms—or miss a great number of detections - when deployed in a different environment [16].
- Expense of Infrastructure and Maintenance: When implementing a network of smart cameras, edge processors, and the necessary cloud resources there will be a significant commitment of infrastructure and operational costs. Each organisation needs to determine what the costs are, and consider these costs in line with the benefits of automating functions, which may also be a perceived barrier to implementing, particularly for small organisations or those with restricted budgets.
- Real Time Processing: Surveillance applications require real time or close to real time processing, and deep learning models, with the complexities an object such a large and technically complex model has, can be expensive on compute; or, a model may require more compute power, or local data than can be used for inference on an edge device. Various forms of solution, from distributed computing to hybrid devices, typically involve designs that utilize cloud edge computing to provide light weight processing on an edge device and heavy.
- Scalability: To scale surveillance systems that cover a significant space with a large number of cameras, need to work with and coordinate a wide range of AI models with high constant data streams. Ensuring that accuracy is consistent throughout the technical scope of the systems remains a difficult task [17].
- False Positives and Negatives: No object detection system is outdated. A false positive (incorrectly detecting a threat) can squander significant security resources and unnecessary alarms, while a false negative (missing a genuine threat) can have dangerous ramifications. There is a continuous challenge to anticipate the precision of the model without discounting the recall [18].

- Adaptability to Changing Environments: Of course, surveillance scenes change with new construction, seasonal changes, new crowd behavior. The models must continually be adaptable or retrained with constant data recollection and annotation.
- Interoperability and Integration Challenge: Many surveillance systems comprise a mixture of classic hardware and software. Employing deep learning models within existing infrastructures, within various IoT devices, and cloud integrations from third-party vendors can be tricky from a technical perspective necessarily needing various protocols and API's to be standardized.

5.3 Medical Imaging

Medical imaging is a vital component of today's healthcare, enabling health care professionals to visualize the structural and functional details of the interior of the human body. X-rays, computed tomography (CT), magnetic resonance imaging (MRI), ultrasound imaging, and positron emission tomography (PET) provide critical information for diagnosis, planning therapies or monitoring therapy. Recently, deep learning technology with applications of object detection have driven extraordinary advances to radiology. Deep learning technologies are showing greater accuracy, efficiency and automated interpretation of medical images.

Deep learning, such as convolutional neural networks (CNN), are able to learn hierarchically composed features from raw images, and can automate the detection, classification, or segmentation of abnormalities found in medical images. Object detection in medical imaging refers to the algorithm learning to understand where an object locates the object and identifies it, such as lesions or tumors, organs, or other pathological structures. Auntomating these processes will help alleviate the demands imposed on radiologists, with more work being drawn from a larger pool to create superior quality diagnoses, generates reduced turnaround times for expedited diagnoses of potential early disease. Figure 5.3 illustrates the applications of deep learning in medical image processing.

5.3.1 Key Applications of Deep Learning and Object Detection in Medical Imaging

1. Tumor Detection and Classification: Perhaps the largest applications in oncology are the detection of tumors and their classification using deep learning models using CT, MRI, or mammography images. CNNs are used during breast cancer screening to identify suspicious masses or microcalcifications. The final models point out suspicious malignancies to radiologists, thereby increasing the sensitivity and specificity of diagnosis [19].

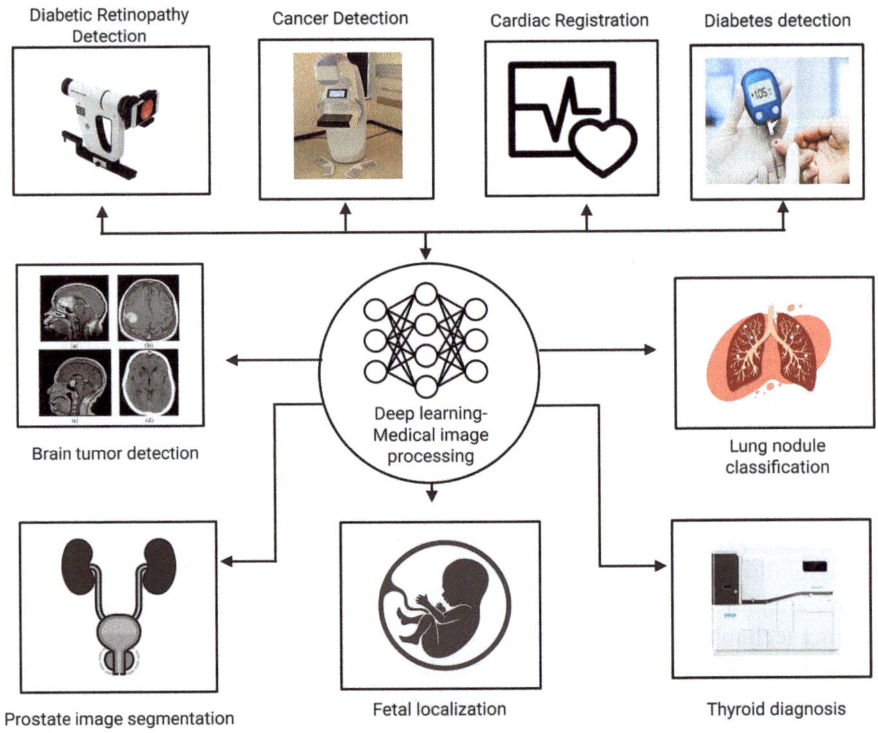

Fig. 5.3 Applications of deep learning in medical image processing. The diagram highlights various use cases, including diabetic retinopathy detection, cancer detection, brain tumor analysis, and organ segmentation, demonstrating the technology's broad impact on healthcare diagnostics

2. Lung Nodule and COVID-19 Detection: During the COVID pandemic, AI systems were built to detect chest opacities of interest on chest X-rays and CT scans. Object detection algorithms locate nodules or lesions attributed either to COVID-19 or some other disease of the lung, thus expediting triage and treatment decision-making.
3. Differentiate Brain Lesions and Stroke: Deep learning supports in ascertaining ischemic and hemorrhagic stroke lesions from MRI or CT images. While it would ensure fast and accurate determination of stroke areas and timely interventions either thrombolysis or determinate surgery, possibly improving the outcome for the patients.
4. Organ Segmentation and Volume Measurement: The goal of segmenting organs such as the liver, heart, or kidneys with precision is fundamental to treatment planning and is particularly important in radiation therapy treatment planning. This process can be facilitated by object detection and segmentation techniques with the detection of individual organs as well as volumetric assessments and dimensional information defining the shape of the organ for patient-tailored treatment.

5. Diabetic Retinopathy Screening: Deep learning models are used to analyze retinal fundus images for diabetic retinopathy markers such as: Microaneurysms, hemorrhages, and exudates. Availability of powered automated screening modalities, these models will allow earlier error intervention for diabetic retinopathy detection in rural locations where one does not have easy access to specialists [20].
6. Histopathology Image Analytic: Deep learning models are trained within digital pathology to delineate cancer cells and tissue structures across biopsy slides. Object detection can also delineate tumor territories in a way that can inform the expert pathologist for tumor grade evaluation and prognosis.
7. CADI and cardia abnormality detection: Echocardiograms, CT angiography, and MRI scans are processed by AI algorithms to identify cardiac abnormalities including valve defects, arrhythmias, or coronary artery disease. These algorithms strive to improve the level of confidence in a diagnosis, and support non-invasive evaluation of cardiac health.
8. Bone fracture and trauma detection: Deep learning algorithms are able to detect fractures and other types of bone injuries in x-ray images. Automated detection advances the ability of emergency departments to triage the injured and treat in a timely fashion.
9. Prenatal ultrasound imaging: Object detection can help in the identification of fetal structures, and anatomical/functional abnormalities in ultrasound images. If a congenital defect is detected early, obstetrical management and care can be improved. The improvements may lead to better outcomes overall.
10. Automated reporting and workflow: AI systems may not only detect abnormalities, they can generate preliminary reports of the diagnosis. The integration of object detection and natural language processing could help advance the work-flow in a radiology department, and facilitate timely reporting and communication.

5.3.2 Challenges and Limitations in Medical Imaging AI

Despite impressive advances, the implementation of deep learning, and object detection in medical imaging solution still faces challenges mainly in terms of IoT devices, cloud infrastructure, and clinical use.

- Computational Constraints and Edge Devices: Clearly many medical imaging devices (such as portable ultrasound or x-ray machines) will qualify as resource-constrained IoT devices, with limited compute, battery life, etc. Running deep learning inference on these edge devices will require model optimization and compression, both of which can impair accuracy and robustness.
- Data Privacy and Security: Medical images contain incredibly sensitive personal health information. Sending large amounts of data to cloud servers for processing

consequently raises significant concerns regarding privacy and security. Additionally, laws such as HIPAA and GDPR enforce compliance and responsibility for secure data handling (including processing), encryption of data, and access to data.

- Large Data Requirements and Annotation Issues: Deep learning models require large annotated datasets for model training. However, annotating medical images is expensive, time-consuming, and requires a strong level of expertise and knowledge. Having insufficient data available may result in overfitting within a category of images, and therefore poor generalizability in performing predictions across all patient populations or imaging devices.
- Variability in Imaging Modalities/Protocols: Performance of a model trained with a dataset from one hospital may not be robust enough to perform well with images obtained from another hospital, because variability is introduced by the imaging machine, acquisition protocol, and patient positioning. Consequently, domain adaptation will be required.
- Integration with Clinical Workflows: AI systems must integrate into existing hospital information systems (HIS), picture archiving and communication systems (PACS), and radiology workflows. Poor integration can limit adoption of a performant algorithm.
- Interpretability and Trust :Transparency and interpretability are required for medical decision making. Black-box deep learning systems can be viewed with skepticism by clinicians. The development of explainable AI (XAI) methods is essential for gaining trust and acceptance.
- Latency and Real-Time Processing: Some clinical applications, e.g., intraoperative imaging, or diagnosis of an emergency, necessitate real-time and/or near-real-time analysis. Inference latency for deep learning, especially using cloud computing for processing, can impact or slow workflow.
- Cost and Infrastructure Requirements: Instating AI (or AI powered) medical imaging systems typically comes with costs for hardware upgrading and setup, and ongoing fees for cloud services, and maintenance. Institutions in resource-limited healthcare environments may face barrier to cost like these.
- Regulatory and Ethical Considerations: All medical AI systems face rigorous clinical validation and regulatory approval processes prior to use in clinical applications. Ethical considerations including bias in training data resulting in health disparities need to be addressed responsibly.
- Reliability and Error Handling: False negatives and false positives have serious implications. False positives can cause unnecessary psychological and emotional stress as well leading to extra imaging tests. False negatives can also lead to delayed interventions and critical treatment. Therefore, there is a continued need for monitoring and to validate AI systems in order to maintain reliability.

5.3.3 Role of IoT and Cloud Computing

The meeting of IoT and cloud computing is important in delivering deep learning clinical applications for medical imaging. For example, an IoT imaging device can capture and transmit, while cloud computing offers scale for training and inference.

- Edge for Low Latency Inference: For time-sensitive applications, reducing latency by processing data using edge/local servers is also a consideration. Directly on-device inference from deep learning models can be implemented with lightweight models, while processing more complex tasks is typically reliant on cloud computing resources.
- Cloud for Model Training and Storage: In the cloud computing infrastructure we can utilize large datasets, powerful GPUs, and in addition to training we can build state-of-the-art models. Cloud computing can also help promote data storage and sharing across institutions and foster collaborative research.
- Federated Learning for Patient Data Privacy: Federated learning allows AI models to be trained in many hospital settings without moving raw patient data, thereby preserving patient privacy, while improving model generalizability.

5.4 Robotics

Deep learning and object detection are improving the capabilities of robotics. These technologies expand the ability of robots to perceive the world, allowing them to execute tasks that require pre understanding of, and then recognizing and responding to multiple objects, in real time. The combination of deep learning, object detection and robotics is transforming the industry in multiple sectors and ushering in a new level of automation completely.

5.4.1 Enabling Perception: How Deep Learning Powers Robot Vision

Robotic intelligence is fundamentally based on perception: the ability to derive meaning from data collected using sensors. While various deep learning algorithms are available, deep convolutional neural networks (CNNs) in particular enable robots to extract meaning from large quantities of visual data. Object detection architectures, such as YOLO, SSD, and Faster R-CNN, enable robots to identify and locate multiple objects in images or from video streams. Essentially this gives a robot visual intelligence that makes perception responsive and adaptive to complex and dynamic environments, for use in decision making and interaction.

5.4.2 *Applications Spotlight: Key Areas Where Robotics Benefits from Object Detection*

1. Accurate Assembly and Defect Detection: Visual inspection systems that robots use to detect defects or imperfections in manufactured parts. Deep learning also takes advantage of the accuracy of deep learning that learns the complex patterns so robots can identify whether manufactured items meet the conditions of use without inspection by humans.
2. Autonomous logistics and warehouse management: Robots in busy warehouses navigate around obstacles, shelving units, and packages by using object detection to improve precision and efficiency within the inventory management workflow: these products get ordered and packaged faster than humans.
3. Healthcare and robotic surgery: Robots assist healthcare workers with simple functions such as recognizing anatomical structures or retrieving access to surgical robots to provide better precision in accurate minimally invasive procedures.
4. Agricultural Automation and crop monitoring: Robots can use object detection to identify ripe fruits and weeds inside agricultural crops. Object detection gives operators the ability to support selective harvesting and target pest management action, while at the same time yields higher crop production.
5. Disaster response and search and rescue: In disaster response categories, it is the case that robotic systems are scanning the debris to determine if there is a person or hazardous object underneath. By utilizing deep learning models robots can help prioritize tasks related to the search and rescue based on determining if a person or object is relevant to the tasks at hand. Figure 5.4 shows a robotic rescue system. The diagram shows a CCTV camera for monitoring, a Jetson Orin embedded AI computer for processing, and a robot designed to assist or rescue a swimmer, illustrating an application in automated safety and response.

5.4.3 *Challenges in Robotics Object Detection*

Despite the shifts in popular perception and consciousness, using deep learning methods for object detection in robotics remains a challenge.

- Limited processing power: Many robots have inherent limitations to the capabilities of their CPU/GPU on-board which limit the types of models that can run on-board. Compression or lightweight models had to be developed to meet existing requirements.
- Variability in the overall environment: Environmental factors such as lighting, occlusions, cluttered backgrounds, etc. will detract from generalization of detection capabilities for operational conditions unique to the laboratory.

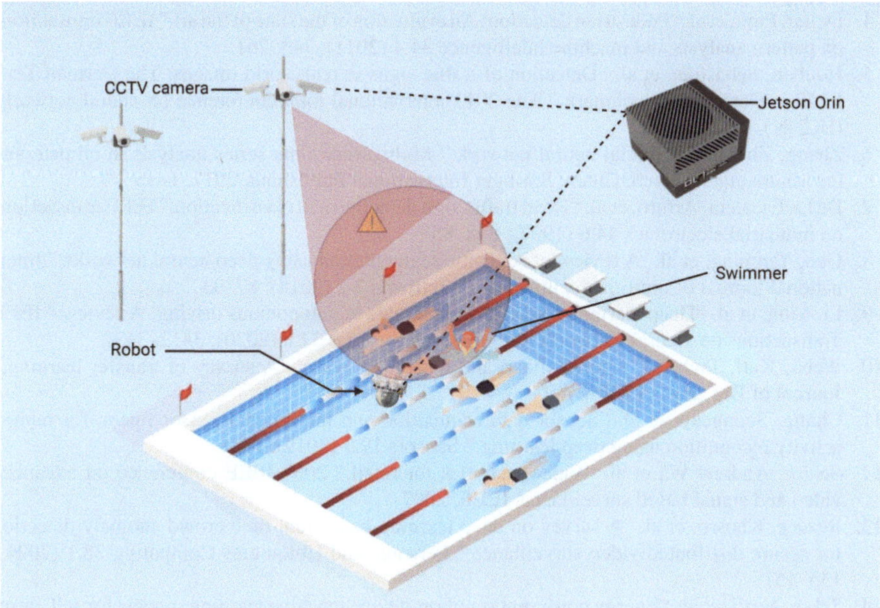

Fig. 5.4 A robotic rescue system. The diagram shows a CCTV camera for monitoring, a Jetson Orin embedded AI computer for processing, and a robot designed to assist or rescue a swimmer, illustrating an application in automated safety and response

- Limited data and costs of annotation: Collecting a wide ranging, annotated dataset is a challenging, expensive and sometimes impossible task that satisfies all eventualities that are likely in a mission-based, or specialized robot task.
- Real-time applications: Robotics applications typically require a new state or action in an immediate time frame as they are directly operationalized in the application. For a system that depends on a complex model and range of operation functions, some tolerance for latency should be agreed upon for returning information related to a decision error that could cause complete failure in an application, designed with the context of related human factors.

References

1. Kumar, Ashwani, and Sonam Srivastava. "Object detection system based on convolution neural networks using single shot multi-box detector." Procedia Computer Science 171 (2020): 2610–2617
2. Yurtsever, Ekim, et al. "A survey of autonomous driving: Common practices and emerging technologies." IEEE access 8 (2020): 58443–58469
3. Müller, Matthias, et al. "Sim4cv: A photo-realistic simulator for computer vision applications." International Journal of Computer Vision 126.9 (2018): 902–919.

4. Dollar, Piotr, et al. "Pedestrian detection: An evaluation of the state of the art." IEEE transactions on pattern analysis and machine intelligence 34.4 (2011): 743–761.
5. Houben, Sebastian, et al. "Detection of traffic signs in real-world images: The German Traffic Sign Detection Benchmark." The 2013 international joint conference on neural networks (IJCNN). Ieee, 2013
6. Zhang, Zhihua. "Artificial neural network." Multivariate time series analysis in climate and environmental research. Cham: Springer International Publishing, 2017. 1–35
7. De La Escalera, Arturo, et al. "Road traffic sign detection and classification." IEEE transactions on industrial electronics 44.6 (1997): 848–859
8. Guo, Yanming, et al. "A review of semantic segmentation using deep neural networks." International journal of multimedia information retrieval 7.2 (2018): 87–93
9. Li, Ying, et al. "Deep learning for lidar point clouds in autonomous driving: A review." IEEE Transactions on Neural Networks and Learning Systems 32.8 (2020): 3412–3432
10. Weiss, Karl, Taghi M. Khoshgoftaar, and DingDing Wang. "A survey of transfer learning." Journal of Big data 3.1 (2016): 9
11. Chung, Seungeun, et al. "Sensor data acquisition and multimodal sensor fusion for human activity recognition using deep learning." Sensors 19.7 (2019): 1716
12. Senior, Andrew W., et al. "Video analytics for retail." 2007 IEEE conference on advanced video and signal based surveillance. IEEE, 2007
13. Rezaee, Khosro, et al. "A survey on deep learning-based real-time crowd anomaly detection for secure distributed video surveillance." Personal and Ubiquitous Computing 28.1 (2024): 135–151
14. Tekin, Nazli, et al. "Energy consumption of on-device machine learning models for IoT intrusion detection." Internet of Things 21 (2023): 100670
15. Veeraiah, Duggineni, et al. "Detection of malicious cloud bandwidth consumption in cloud computing using machine learning techniques." Computational Intelligence and Neuroscience 2022.1 (2022): 4003403
16. Jadhav, Arya, Suhila Sawesi, and Bushra Rashrash. "Bias and generalizability challenges in machine learning models for leptospirosis." 2024 IEEE International Conference on Big Data (BigData). IEEE, 2024
17. Coates, Adam, et al. "Scalable learning for object detection with GPU hardware." 2009 IEEE/RSJ International Conference on Intelligent Robots and Systems. IEEE, 2009
18. Zhao, Zhong-Qiu, et al. "Object detection with deep learning: A review." IEEE transactions on neural networks and learning systems 30.11 (2019): 3212–3232
19. Sajiv, G., and G. Ramkumar. "Multiple Class Breast Cancer Detection Method Based on Deep Learning and MIRRCNN Model." 2022 International Conference on Inventive Computation Technologies (ICICT). IEEE, 2022
20. Li, Tao, et al. "Diagnostic assessment of deep learning algorithms for diabetic retinopathy screening." Information Sciences 501 (2019): 511–522

Chapter 6
Deep Learning Frameworks for Object Detection

Abstract This chapter offers an in-depth examination of the frameworks and tools related to deep learning that are important for modern object detection. The primary focus will be on the two most prevalent ecosystems–TensorFlow and PyTorch–and the common strengths for deployment to production and flexibility for research. Other specialized libraries, such as Detectron2 and OpenVINO, are discussed as they support the object detection development process by simplifying model development and optimizing inference. This chapter builds onto the practical section regarding the value of pre-trained or transfer learning models which enable high performance using minimal data, and also highlights important strategies for deploying in real-world settings such as model compression techniques (e.g. pruning and quantization) for deployment on edge devices, and cloud deployment strategies for serving any data. With this discussion, practitioners will understand when the models and accompanying tools/strategies can be integrated for building effective object detection systems in a range of situations from embedded devices to cloud-based infrastructure.

Modern object detection workflows are based on general-purpose deep learning ecosystems. In the ecosystem landscape, there are two major players, TensorFlow and PyTorch, which are the two most used frameworks. TensorFlow, developed by Google Brain, has a massive ecosystem and has pioneered many tools for building, training and deploying models from edge devices to data centres; even APIs for specialized deployments that facilitate deployment pipelines . PyTorch, developed by Meta AI, has aimed towards the most natural, Pythonic interface with dynamic computation graphs, and a quality developer experience—characteristics that won it the support of academics and industry-level experiments. That being said, PyTorch has debugging and iterative model design features, which are also claimed advantages over TensorFlow [1].

In addition to these foundational libraries, there are specialized toolkits that focus specifically on object detection and computer vision tasks. Various examples exist, such as Detectron2, which is built on top of PyTorch. Detectron2 is a modular, production-grade vision library developed for object detection workflows, contains

A. Elhanashi and S. Saponara, *Deep Learning for Object Detection and Localization*,
https://doi.org/10.1007/978-981-95-4846-0_6

a lot of community support and documentation, and can serve as either a starting point for a new vision library or as an end-to-end detection pipeline [2]. Similarly, PaddlePaddle the deep learning framework developed by Baidu has a dedicated detection package that simplified the entire training, evaluation, and deployment pipelines in real scenarios, especially in regards to applying the models to industry and autonomous systems [3].

Often times, applying detection performance to real world scenarios requires significant consideration of model deployment and run-time optimization. There are frameworks such as OpenVINO (developed by Intel) which offer solid methods to optimise neural networks for inference, especially on CPUS, integrated GPUs and essential edge device [4]. These tools apply model-level graph transformations, reductions in precision and hardware specific optimizations, assisting developers to obtain low-latency resource-efficient performance.

AutoML frameworks are also transforming how object detection systems are created. Examples include NVIDIA TAO and AutoGluon, both of which try to automate the preprocessing of datasets, hyperparameter optimization, and architecture selection to lower the threshold for non-experts [5]. Google's Vertex AI encompasses this approach by providing a single interface for managing the complete model lifecycle, from training to deploying, across multiple data modalities. Each platform condenses the detection workflow into more easily-consumed pipelines while abstracting much of the manual engineering historically associated with them.

Importantly, frameworks like OpenCV are still relevant, and used, in object detection pipelines. The basic use of the framework is rooted in classical vision approaches (with potential for other approaches), but it also supports loading and running your own additional deep learning models and even provides GPU acceleration support for pre- and post-processing of real-time images. This capacity allows for the user of neural networks into traditional CV. Figure 6.1 shows the ecosystem of major deep learning frameworks and their primary corporate supporters.

6.1 TensorFlow and Keras

Launched in November of 2015 and built by Google Brain, TensorFlow is an open source library for building, evaluating, and using machine learning models across a wide array of hardware (CPUs, and GPUs, or mobile devices or edge device deployments). TensorFlow's architecture is based on dataflow graphs (with "tensors" flowing through computational "ops"), but can scale to multiple machines or accelerated hardware such as TPUs. It implements many advanced features—including symbolic graph mode and eager execution. This allows you to develop a model via execution in an albeit limited manner, with your model's processing graph hooked together dynamically, thus allowing rapid iterations and easy debugging. It includes built in auto differentiation elevators which simplify the process of backpropagation by using automated differentiation techniques to autodiff gradients while you are training your models [6, 7].

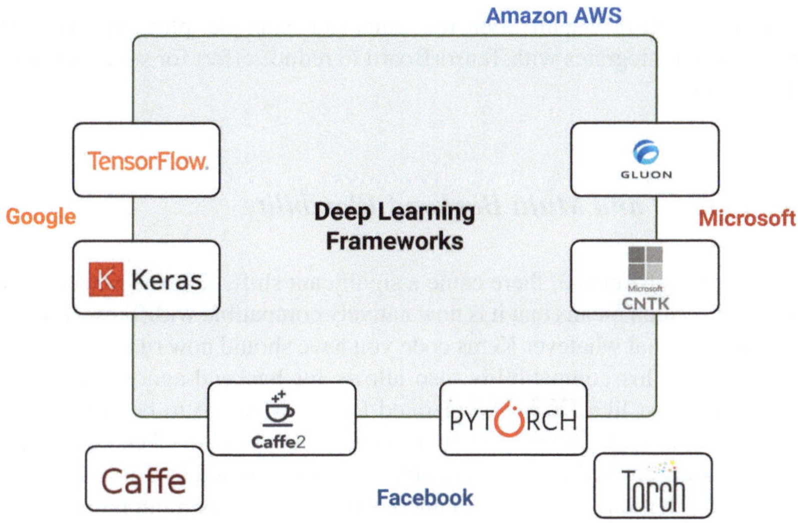

Fig. 6.1 The ecosystem of major deep learning frameworks and their primary corporate supporters, including TensorFlow (Google), PyTorch (Facebook), Keras, Caffe, and their integration with cloud platforms like Amazon AWS and Microsoft Azure

TensorFlow has a large ecosystem:

- TF Lite for on-device/mobile inference
- TF.js to run models in the browser
- TF Extended (TFX) for enterprise pipelines
- TF Hub for re-usable pre-trained components

These established capabilities democratically support the complete life cycle of the model development (collection, training, deployment and scaling,) across varied environments and teams. Its cross platform nature has propelled numerousQuotes its AI services that's powered a suite of products and services including and AI powered products in Google Photos, Google Translate, maroc and in some cloud-based AI products.

Keras was first developed in 2015 by François Chollet [8] as an easy-to-use Python API for neural networks that runs on various backends (first TensorFlow and then Theano and CNTK). Since it was modular and human-centric, it became preferred for quick prototyping and obviousness Developers had the choice to make use of the Sequential API—easy for simple, linear stacks—or the Functional and subclassing APIs, which allow for custom architectures and advanced workflows Keras is focused on clarity, modularity, and developer productivity Just mix and match layers, optimizers, and losses with little code, and use well-defined standard data structures (such as NumPy arrays, Pandas dataframes, or TensorFlow/PyTorch data loaders) Keras also provides built-in visualization functions (for example, plot_models or model summaries) and integrates with TensorBoard to reduce effort for some development and debugging.

6.1.1 *Keras 3 and Multi Backend Flexibility*

With the arrival of Keras 3, there came a significant shift. Keras is now a real multi backend API, which means that it is now natively compatible with TensorFlow, JAX, and PyTorch, so that whatever Keras code you have should now run on all three execution engines. This compatibility also allows for backend-agnostic performance optimizations must like XLA (Accelerated Linear Algebra) to be utilized without having to rewrite code. Keras 3 also brought in new cross-backend compatibility, such that models created in Keras are able to be instantiated as TensorFlow SavedModels, PyTorch modules, or JAX functions to give the maximum possible flexibility for deployment. In addition, keras_core.ops now uses a unified keras_core.ops namespace, which allows writing custom operations that will work in all supported frameworks.

6.2 PyTorch

At its core, PyTorch's computation graph is dynamic—its graph is built at runtime instead of requiring a prebuilt graph like other frameworks [9]. This gives excellent intuition when debugging, and allows for fast experimentation: developers are free to just use Python flow and control structures. A post from one user in Reddit said, "PyTorch just works the right way with its very flexible nature, and it's feasible to easily use it in production." The operations in PyTorch are all based on tensors—multi-dimensional array structures like NumPy arrays, that can be run in parallel do to their propagation on GPUs. The built-in autograd will keep track of tensor operations automatically in order to set up gradient calculations, enabling efficient backpropagation without the need for the user to perform manual differentiation tomorrowdesk.com+1en.wikipedia.org+1. Automatic gradient tracking also provides seamless support of in-place, memory calculation optimizations when training complex architectures. As a library aimed at research, developers have access to a rich catalog of neural elements used to compose architectures, including convolutions, recurrent layers, activation functions, and loss functions, while torch.optim provides commonly used optimizers such as SGD, Adam, and RMSProp tomorrowdesk.com. Automatic Mixed Precision training (AMP) allows for mixed-precision training, further lowering memory consumption and increasing GPU performance. TorchScript narrows the gap between research and production, allowing one to convert fairly complex dynamic PyTorch models into static, serialized graphs that are independent of the Python runtime. This enables efficient deployment of the model on hardware configurations such as servers, mobile devices, or embedded systems, following one of two paths: tracing or scripting. The PyTorch ecosystem contains full-featured dedicated libraries:

- Torchvision (for computer vision)
- Torchtext (for natural language)
- Torchaudio (for audio)

It also works with ONNX [10] so you can export models and make them available across frameworks. One community member stated, "Pytorch Lightning does bring a cool modularity vibe... experimenting is just less about the model per se... It quickly becomes a mess." Frameworks like Pytorch Lightning make full-stack training easier by abstracting the boilerplate, while still allowing for scalability across GPUs and clusters datasincesbase.com. Furthermore, tools like DeepSpeed, and accelerator plugins (e.g. Hydro, Accelerate) are pushing the limits of the performance with techniques of memory sharding and other distribute strategies.

6.3 MATLAB

MATLAB, a proprietary numerical computing environment and programming language, was developed by MathWorks, and initially released in 1984 MATLAB is a multi-paradigm numerical analysis, simulation, algorithm development, and data visualization tool It has an easy-user interface and is attractive to diverse institutions including private, academic (there are thousands of universities in the world), and governmental entities that may be in industries, such as aerospace, engineering, or finance, industries that became targets in happy years for software.

The Deep Learning Toolbox sits at the core of MATLAB's deep learning capability [11], with functions to design, train, analyze, and deploy models. This toolbox contains both programmatic functions and GUIs that have less programming, e.g. the Deep Network Designer and Experiment Manager Users can design sophisticated architectures using convolutional networks (CNNs), recurrent networks (RNNs), and for other architectures such as transformers long the process, this toolbox gives users visualization and layer inspection while designing their models.

The Deep Learning Toolbox framework also supports model optimization; users could build their networks and do quantization, pruning, and projections to make their networks leaner and ready for deployment.

MATLAB simplifies the complete data-to-deployment process. For larger datasets, datastores provide efficient model iteration without exceeding memory limits while importing the files. These interactive apps—Image Labeler and Video Labeler—help users annotate training data with greater precision. After prototype design and model training, MATLAB can deploy to several targets: GPUs, CPUs, embedded hardware, FPGAs, and generate C/C++, CUDA, or HDL code. The ability to integrate designs with Simulink allows users to model an entire system—neural models can be simulated at system-level within a realization control or signal-processing pipeline. This will be crucial for embedded systems and when considering cyber-physical systems.

MATLAB fosters collaboration across technol stack. It supports importing/exporting several proprietary (but open) aspects: ONNX, Keras/TensorFlow™, and PyTorch®, including making calls within Python packages, and Model exchange with Matlab→Python via ONNX. Also, Matlab's Model Hub provides access to current popular pretrained networks—they are 100% ready to for transfer learning.

Finally, MATLAB's low-barrier entry is enhanced by the productivity of its apps. The Deep Network Designer enables users to visually connect layers into architectures, and offers a graphical way of viewing the layer details while training—cool! The Experiment Manager app facilitates viewing variation in trials from across hyperparameters, and the labeling apps facilitate efficient generation of training ground-truth data.

As someone who using MATLAB, what I like about it is that "you don't have to deal with packages." When performing research and/or industrial prototyping—especially for embedded systems—MATLAB is still a powerful environment for prototyping and development. Hardware connections/flows from MATLAB to simulation tools (Simulink, HDL Coder, GPU Coder), allow workflows to maintain system-model consistency and/or real-time performance. However, in the wider machine learning ecosystem, many communities of developers/grokking researchers are using Python-based tools. MATLAB is a great environment for quickly prototyping models, producing workflows for complex integrated-system design (with engineers), and when workflows continue to align with hardware (especially if you are modelling and executing in engineering). Closely integrated with its Deep Learning Toolbox, MATLAB has been adding interactive apps, expanding the number of deployment targets, simulation workflows, and allowing developers and researchers to create embedded and applied AI workflows. MATLAB's reliance on a closed ecosystem, and its associated potential costs of licensing and environment in relation to the human-object and design in competition with open-source Python-based packages, limit its space for developing state-of-the-art research or large-scale production. In summary, MATLAB is a powerful engineering-focused ecosystem that takes you from data labeling to deployment in a seamless experience, where it excels in spaces in engineering that have potential to interface with a physical system. For developing bleeding-edge model innovations, or deploying scaled models in industrial production spaces, python-based tools are currently leading the way.

6.4 OpenCV

OpenCV [12] (Open Source Computer Vision Library) is a powerful, open-source solution that brings together traditional computer vision functions with cutting-edge deep learning features. Released under the permissive Apache 2.0 license, OpenCV is written in C++ and has its own bindings for Python, Java, JavaScript, MATLAB, etc. This makes it quite a portable solution. The library began in the late 1990s at

Intel, where it later became modified by Itseez (which is now part of Intel). The main objective of OpenCV is to facilitate real-time computer vision. The first stable OpenCV release happened in 2006. The library was updated with major versions 2.x in 2009, and 3.x in 2011 with the overall focus on performance, safety, and usability increases. New updates have become common place: although it still makes many users happy when a major new version comes out, magazine subscriptions and released first editions cannot beat the reality of consistent open-source updates like the ones that OpenCV releases personally. OpenCV is updated and released about twice/year ensuring its ability to operate under the current hardware compliance/testing standards. Today, OpenCV contains over 2,500 optimized algorithms. In today's OpenCV you will find everything from basic image and video I/O, filtering, morphoogical ops, geometric transformations, feature detection (e.g., SIFT, HOG, Harris detector, LBP), object tracking, stitching, camera calibration, structure-from-motion, and 2D/3D reconstruction.

The modular nature of OpenCV, with over 70 core and contrib modules (e.g., core, imgproc, video, feature2d, objdetect, rgbd, and optflow) is enhanced by contrib modules on structured light, super-resolution, augmented reality markers like ArUco etc. The library since version 3.1 has included a DNN module, which permits deep neural network inferencing without requiring Frameworks such as tensorflow and pytorch Since version 3.3 this was promoted to the main repository, with full ONNX and Frameworks support via extensible backends for portable and accelerated inferences that include Intel's OpenVINO, NVIDIA cuDNN, and Vulkan support. The capabilities of DNN module is expanded in OpenCV version 4.11 released in 2025 with on-deman blockwise quantization, support for a wider range of ONNX functionality, HAL support for Qualcomm and RISC-V, and improved GPU memory usage which lowers inference time and is useful for memory restricted products. Real-time inferencing tasks, including detection and segmentation, and classification tasks deployed on constrained hardware can be undertaken with OpenCV. OpenCV is highly optimized for working with video streams and camera feeds, and it is capable of deploying GPU-accelerated functions for common processes such as resize, denoise, color conversion, and format transcoding. OpenCV works with almost any deep-learning Framework because of ONNX model support, and it is capable of pre-loading trained networks in an inference pipeline. The library provides the ability to build and prototype computer vision systems in real-time particularly for applications in robotics, medical imaging, surveillance, UAV's etc. The library supports cross-platform deployments: Windows, Linux and macOS, both Android and iOS mobile platforms, embedded platforms like QNX, and even web applications using OpenCV. With wrappers available in many languages, OpenCV can be built into varied production stacks or research environments. OpenCV has an ecosystem developed by the OpenCV Foundation, which provides documentation, tutorials, forums, courses, and events. Resources include official documentation, community sources such as LearnOpenCV, and companion code to "Learning OpenCV 4" published by Packt. The annual "Perception Challenge" invites developers to define the limits of vision-based AI systems. Figure 6.2 illustrates the main modules of the OpenCV library,

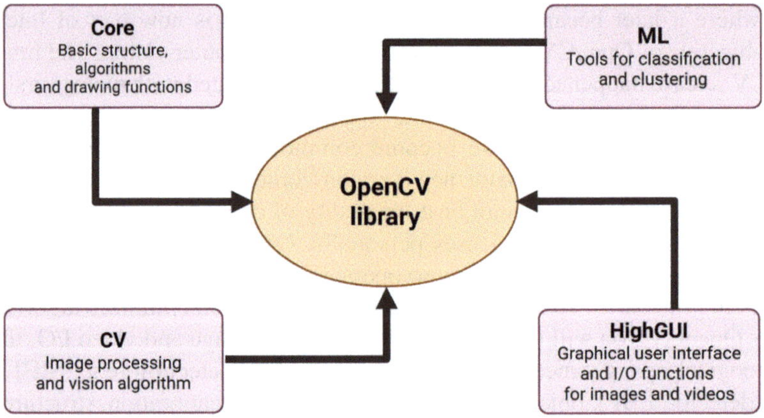

Fig. 6.2 The main modules of the OpenCV library, including the Core (basic structures and algorithms), CV (image processing and computer vision algorithms), ML (machine learning tools), and HighGUI (graphical interface and I/O functions)

including the Core (basic structures and algorithms), CV (image processing and computer vision algorithms), ML (machine learning tools), and HighGUI (graphical interface and I/O functions).

6.4.1 Strengths and Limitations

Strength

- A big, optimized library of core vision algorithms—efficient, portable, and time-tested.
- Hardware acceleration using CUDA, Vulkan, OpenCL, and OpenVINO.
- Applicability to deep learning inference that is flexible and independent of a framework.
- Flexible platform and language support means quick prototyping and production deployment.

Limitations:

- The DNN module can only be used for inference, other frameworks are required for training networks.
- Less support for research phase innovation that typically exists in Python based AI ecosystems.
- The model zoo is smaller than TensorFlow Hub, or the PyTorch repository.

6.5 Caffe

Caffe [13] (Convolutional Architecture for Fast Feature Embedding), the brainchild of Yangqing Jia at UC Berkeley's Vision and Learning Center (BVLC), was announced as an open-source project in late 2013. Caffe was the right thing at the right time, when efficient and fast processing of image data was becoming important, especially with the explosion of ImageNet research. Caffe's C++ core with Python and MATLAB bindings for efficient training or deployment of deep models with convolutional neural networks was built from the ground-up for using GPUs. There are three specific principles—speed, modularity, and expressiveness—that drive the adoption of Caffe. Caffe is declarative in nature, meaning architectures, solvers, and flows of data are defined in plain-text configuration files (protobuf-based) so it is easy to experiment with various model configurations. In its simplest form, a model's flows can be moved between CPU and GPU with a toggle of a flag, thus enabling rapid prototyping. Benchmarks show up to 60 million images each day on a single NVidia K40, about 1 ms per image for inference. Caffe's modular layer-based architecture is designed for developers to create different architectures, such as convolutional, fully connected, recurrent, and more. Caffe does this through a core "net", "layer", and "blob". New layers or new loss functions are possible through C++/CUDA. C++/CUDA are a lower-level coding language that allows extensibility. The C++ performance and Python/MATLAB availability of Caffe built a community that was a helpful asset to researchers. The official Model Zoo had links to pre-trained networks to do transfer learning that have shorter experimental cycles. Other programs, such as Barista, GUI-driven ancillary programs for Caffe, made it easy to design the network. Academic researchers began to adopt Caffe as a tool for vision-related research tasks (e.g., classification, segmentation, detection), not uncommon given the fast training routines and ease of use. It became the choice of tool in fields, many academic laboratories, and production pipelines at startups and large enterprises. CaffeOnSpark expanded Caffe to operate on Apache Spark, and within a Hadoop cluster, enabling distributed training on-premise without needing to move the large datasets off-prem. Caffe with Troll (CcT) offered some CPU-batching as end user experience, and optimized different things enhancing performance in numerous situations. In 2017, Facebook launched Caffe2 to tackle more general architectures and distributed situations; it even enabled mixed-precision on NVIDIA Tensor Cores. However, by 2018, Caffe2, and PyTorch were merged back into a unified PyTorch ecosystem.

6.5.1 Strengths and Limitations

Strengths:

- Training and inference at lightning speed.
- Separation of functionality utilizing cleaned, modular architecture defined via configuration.
- Toggling between the GPU and CPU is straightforward and deployment can scale where necessary.
- Mature ecosystem with pretrained models and community support.

Limitations:

- In many cases, adding novel layers or architectures requires C++/CUDA coding
- Not as nimble as several newer frameworks (ex. PyTorch) for research-led innovation
- Community engagement and development slowed after 2017; officially support is largely in maintenance mode.

6.6 Using Pre-trained Models

Pre trained models provide a very strong start by providing pre learned features from large and often diverse datasets. It's hard to train from scratch without the right dataset, and it often requires large compute resources or cloud computing to even be practical. Pre trained models give an option, as they come ready with general features such as edges, textures, and shapes and can quickly be fine tuned without many of the possible new tasks with a much smaller dataset. A large reduction in training time and infrastructure cost is a path to quick prototyping. Figure 6.3 shows the comparison of deep learning training strategies: training a model from scratch, using a pre-trained model out-of-the-box (with frozen layers), and fine-tuning a pre-trained model (where some layers are frozen and others are trained).

6.6.1 Fine Tuning Versus Feature Extraction

There are two general strategies that most likely apply when using pre trained models:

1. Feature Extraction: Freeze the early layers, which often capture low level patterns, and train only the latter layers or classifier head. This can be very efficient and will work quite well if the differences between domains are moderate.

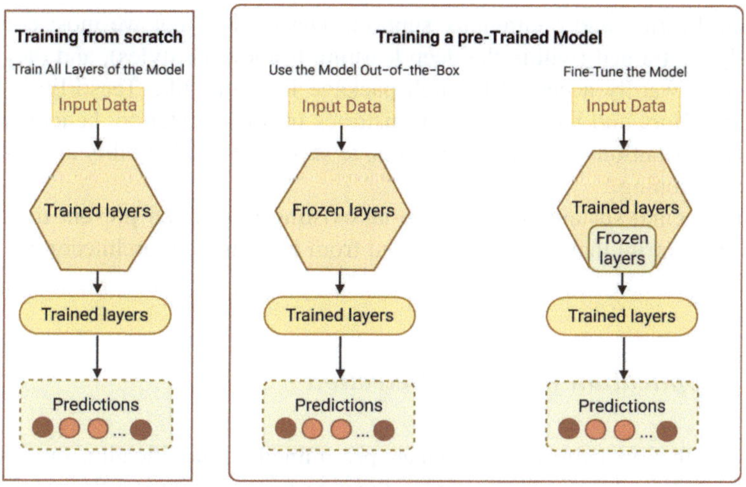

Fig. 6.3 A comparison of deep learning training strategies: training a model from scratch, using a pre-trained model out-of-the-box (with frozen layers), and fine-tuning a pre-trained model (where some layers are frozen and others are trained)

2. Fine Tuning: Unfreeze and jointly train multiple layers, and sometimes the entire network, to better align with the specific dataset. This method may require more resources, but typically can provide more accuracy, and preferable when there is enough labeled data.

6.6.2 Advantages of Pre Trained Models

- Data efficiency: In low-data operating conditions—like with medical imaging or specialized domains (like text classification or some types of climate predictions)—pre-trained models can be data efficient due to being able to transfer learned features associated with some generalized forms of domains (e.g., ImageNet).
- Time and resources: minute by not needing to train from scratch, your team can spend its time concentrating on things like model selection, hyperparameter tuning, and deployment.
- Performance: Pre-trained architectures also have the potential to provide better generalization and overall accuracy than architecture trained from scratch, especially when done on small to moderately sized datasets, with significant variance in performance compared to other methods as dataset size decreases.
- Robustness and generalization: By being pre-trained on a large number of diverse examples, pre-trained architectures become more resilient against overfitting and can show better performance when encountering data they have never seen before.

- Standardization and community support: The models that we most commonly use all pre-trained (within the deep learning framework styles), and they can be straightforwardly accessed through package locations like TensorFlow Hub or PyTorch Hub, and have significant histories in documentation, benchmarks and also redistribution by community users, to support reproducibility and guideline-based reasoning.
- Environmental sustainability: Using an existing model can prevent the massive carbon footprint that would be awarded from training your architecture.

6.6.3 Application in Object Detection

In object detection, using models that are pre-trained on classification, or even on the detection task itself, can produce immediate performance improvements. Evidence suggests that pre-training on a classification dataset (i.e., ImageNet), or even better, from a large scale detection dataset (COCO, OpenImages), can lead to improvements in sample efficiency, as well as improved localization on smaller domain target datasets. For instance, pre-training on object detection data provided a 7.6% mean average precision (mAP) improvement on PASCAL-VOC over pre-training only on a classification dataset.

6.6.4 Common Challenges and Considerations

- Domain Mismatch: When the domains of the source data and target data are different (e.g., natural versus medical images), the transferred pre-trained features may perform poorly. Therefore, careful domain-aware adaptation will be necessary.
- Overfitting Risk: Applying fine-tuning using a very small dataset carries a risk of overfitting [14] if all the weights of the network are updated without very many examples.
- Model Biases: Pretrained models contain the biases in the training data and may be copies of the training data errors and blind spots.
- Large Sizes: Many types of pretrained architectures consume a large amount of resources. Further, they are often not practical for deployment unless they can be pruned, quantized, or have an effective substituted efficient model.

6.6.5 Best Practices for Fine Tuning

- Freeze Early Layers: Initially freeze shallow layers, adjusting only deeper layers or now-added heads; then unfreeze them slowly as necessary [16].

- Use Lower Learning Rates: Allow learning rates 10–100x smaller than training from scratch to increase stability of adaptation.
- Strong Regularization: Use data augmentation, dropout, or weight decay to manage overfitting while fine-tuning.
- Evaluate Several Backbones: Evaluate a variety of pre-trained architectures (CNNs, Transformers, self-supervised models). Recent evaluations indicate that traditional CNN backbones still perform better than many of the more recent methods in object detection benchmarks.
- Layer Wise Training: Unfreeze blocks of the model one by one to allow lower layers to adapt and limited lower-level adaptations.

6.7 Model Compression and Quantization

Object detection models typically have millions of parameters, and they are usually loaded in 32-bit floating-point values. The precision of 32-bits guarantees accuracy when detecting, it has the disadvantages of being large, Power Hungry, and Slow—all hurdles to deployment on a mobile device, drone, or IoT application. Model compression and quantization are two strategies to circumvent these hurdles, they can make a model smaller and run faster with little adverse relationship to Will help-with detection accuracy which Allows a detector to run on edge hardware. The most intuitive model-compression strategy is pruning, where irrelevant connections, not commonly, are typically zero weights are pruned from the model. When we zero out a weights, the model becomes sparse. If hardware has sparse-matrix capabilities then sparsifying reduces the time to inference. More advanced forms of pruning not only prune individual weights, prune entire filters, channels, and able to limits arrangement structured sparsity which can reduce the computational cost to detection while retaining their accuracy. Quantization reduces number precision, by converting from 32-bit float to lower bit formats like 16-bit float and/or 8-bit integers. This technique can reduce memory usage by up to four times, and even improve inference speed, particularly on low-precision optimized processors. Post-training quantization (PTQ) is a very simple way to quantize a fully trained model. PTQ is simple and fast, since it quantizes an already trained model without retraining, but it does have the cost of losing some accuracy. To solve this accuracy issue, quantization-aware training (QAT) has been developed to pull quantization into the training loop, such that the model can adapt to low precision during training, and often retain a lot of the accuracy. Pruning and quantizing can be performed in sequence; first we can prune redundant parameters, then we can quantize the remaining parameters for further representation efficiency, while maintaining the same detection quality. Researchers have even created joint methods that combine pruning and quantizing, where you can determine the bit width and sparsity layer by layer using AutoML, such that we can tune the representation size and detection performance. These pruning and quantizing techniques are particularly critical for object detection, where we must draw bounding boxes precisely, not just identify classes. Mixed-precision pruning methods can retain dependent localization filters while teacher-student methods like Quantization Mimic leverage model knowledge from a quantized model to increase performance of a lightweight detector. When considering real applications, apparent successes arise: one project incorporated pruning as well as INT8 quantization for an object detector and produced a model size that was twelve times smaller while achieving inference that was as fast as ten times faster on a CPU without impacting mean average precision (mAP) much. Another project applied sparse quantization to YOLO based detectors and produced similar improvements - achieved 14.5x compression, real time laptop CPU inference (around 20 frames per minute), and a 2 mAP point decay in performance. Compression involves trade-price. More impacted formats like 4-bit or binary allow for greater compression but most likely at the cost of more accuracy. Also related to accuracy is how hardware is affected; an INT8 gain is often only achieved on CPUs or accelerators that use vector instructions like VNNI otherwise, we will not likely get the extra speed we assumed with lower precisions. Compression techniques should be based on hardware, application requirements, and needed accuracy.

At a practical level, engineers take a baseline model and examine its size, speed, and accuracy. Then, they prune and quantize (with QAT or teacher-student models recovering from it), put the model on their target hardware, and test its performance and finally tune the compression settings. In practice, good results can usually be achieved by combining channel pruning with INT8 quantization to get a reasonable performance vs size trade off.

In the end, compression and quantization [15] methods compress and quantize large detection models into lightweight, energy-efficient, real-time detection systems for use on smaller hardware. By carefully pruning and stripping-down networks, developers are able to deploy object detectors at scale, easily deploying task-oriented accurate object detectors on everyday hardware like smartphones, drones, and autonomous sensors. Figure 6.4 illusrates the quantization in digital signal processing.

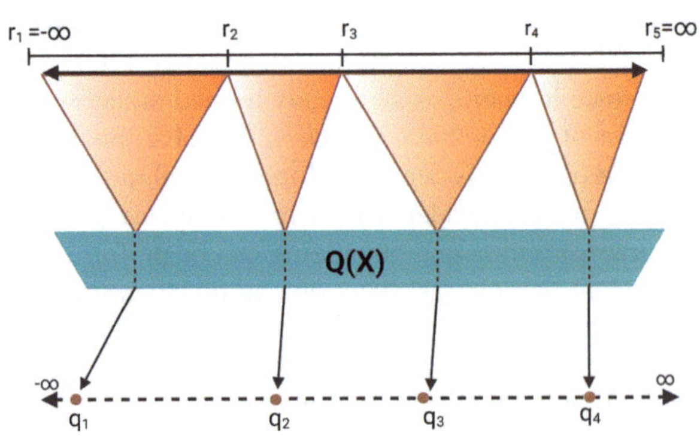

Fig. 6.4 Illustration of quantization in digital signal processing. A continuous signal Q(X) is mapped to discrete quantization levels (q1, q2, ... q1, q2, ...) using a set of thresholds (r1, r2, ... r1, r2, ...)

6.8 Deployment on Edge Devices and Mobile

Running object detection directly on a local device (such as a smart phone, embedded systems, or industrial cameras) has significant advantages: the major benefits are that all inference can happen immediately with performance as fast as the device itself and can't require any internet, now or in the future. There are also privacy concerns, which result in having all raw image data being retained local, meaning a cloud service always doesn't have access to it. Bandwidth also improves from the point of view of reduced fees and delays. Figure 6.5 shows the hardware acceleration support in modern computer vision libraries. The diagram shows how libraries can leverage different processors (CPU, GPU, FPGA) and vendor-specific libraries (like oneDNN, ARM Compute Library) for optimized performance across various platforms.

6.8.1 *Core Constraints and Roadblocks*

Edge and mobile deployment have a few built-in limitations:

1. Compute and Memory Bottlenecks: These devices often do not have data-centre GPUs or TPUs and are limited to considering inexpensive low-power CPUs, NPUs, or microcontroller elements that may only have kilobytes of RAM. Not only can you not store or evaluate large models, but optimized techniques such as pruning, quantization, and knowledge distillation are often needed.

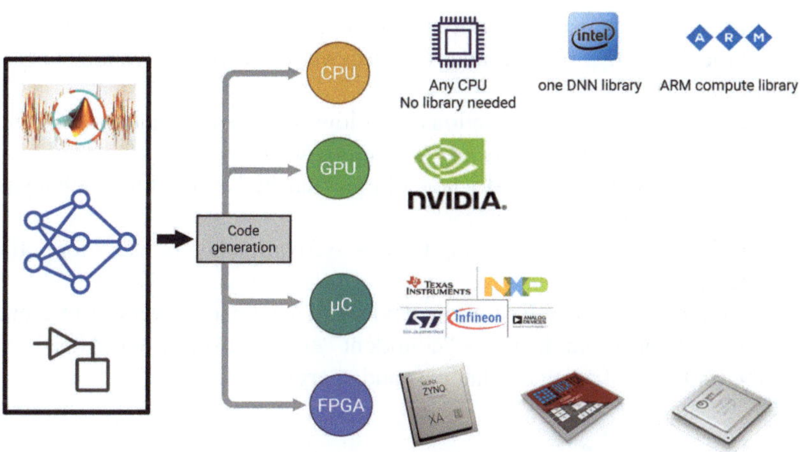

Fig. 6.5 Hardware acceleration support in modern computer vision libraries. The diagram shows how libraries can leverage different processors (CPU, GPU, FPGA) and vendor-specific libraries (like oneDNN, ARM Compute Library) for optimized performance across various platforms

2. Power Consumption and Heat; Battery powered units need to balance performance and by design are energy efficient and can lack out-and-out raw computing power. Not only can CPUs overheat or throttle when fanned storms from sustained load, but mobile OS schedulers will often restrict heavy content to prolong battery life or prevent overheating!
3. Latency Requirements: Robotics, autonomous drones, or AR being a few examples, are unique use-cases that require millisecond-level response times to avoid robot or drone crashes. Offloading and processing data in the cloud with unacceptable latency makes on-device inference essential.
4. Hardware and OS Fragmentation: A mobile environment is a mishmash of ARM CPUs, Mali/Adreno GPUs, NPUs, and special non-ASIC accelerators, all with different quirks and driver issues. Even Android it self, has several versions of Android and also has gpu driver fragmentation of its own which complicates deployment!
5. Storage Limits: Keeping models onboard without downloading to storage is also a problem because even the state-of-the-art detectors can easy run hundreds of megabytes to store without compressing data. Techniques like compression and early-exit models can help offload some of this responsibility.

6.8.2 Optimization Strategies Simplified

Here are essential strategies simplified:

- Quantization, which usually reduces weights from 32-bit floats to 8-bit ints, can shrink models by $75+\%$ but reduce accuracy very little (and often may not reduce accuracy at all).
- Pruning, which removes connections with low importance, reduces the size of a model and the computational need for training or inference.
- Knowledge Distillation is used when a large "teacher" is used to teach a smaller "student" that performs similarly.
- Operator Fusion, in this case fusing Conv + Activation layers with more efficient compute options.
- Split Computing & Early Exit allow devices to send early layers of inference or stop inference once the device is "confident" enough using Look Ahead to help with the trade-off of accuracy, latency, and energy.

6.8.3 Frameworks That Help

With specialized tools, deployment can be simplified:

- TensorFlow Lite (and Micro) supports quantized models across Android, iOS, and embedded MCU.
- ONNX Runtime and OpenVINO, which support heterogeneous hardware acceleration, also make deployment easier.
- PyTorch Mobile, typically used with TorchScript and optimized using the .ptl format on mobile devices.
- Core ML and Metal Performance Shaders (MPS) for iOS; NCNN, MNN and TensorRT for Android.

6.8.4 Best Practices for a Smooth Deployment

1. Choose the right hardware according to performance versus budget: Jetson Nano, Coral TPU, Cortex-M microcontrollers, or NPUs.
2. Optimize models as much as possible with quantization-aware training, pruning, and distillation.
3. Use the most suitable inference engine, aware of the hardware, for the highest performance.
4. Profile under real-world conditions: ambient light, multiple threads, OS load.
5. Being monitoring and updating continuously: for performance improvements and security patches.
6. _model binaries_.safe storage, and the integrity of firmware should be guaranteed in production.

6.8.5 Challenges to Overcome

- It usually remains tough to maintain accuracy after compression, and quantization-aware training might be required.
- Changes in lighting and movement require robustness.
- The OS and hardware diversity in the mobile landscape adds to this complexity.
- It is managing energy consumption against perfomance so as to not get an thernal shut-down.

6.9 Cloud Deployment

In a cloud deployment of object detection, the model is hosted on a remote server (e.g., Amazon Web Services, Google Cloud, or Microsoft Azure) rather than a local device. These platforms have scaleable compute resources (GPUs, dedicated engines, etc.) that free the user from supplying detection services via Rest APIs or streaming in real time. A cloud deployment model is also justified when needing to provide to many clients, high workloads, or to integrate with other cloud services (e.g., storage, monitoring, analytics pipelines, etc.).

6.9.1 Typical Cloud Deployment Workflow

A common workflow is to train a model (whether custom or pre-trained) on cloud infrastructure or locally. Then after you export it out to a portable format like TensorFlow SavedModel, ONNX, or PyTorch TorchScript. Containerize it with Docker, and run your model on a cloud service such as TensorFlow Serving (running inside a Docker container), AWS SageMaker, Google Cloud Run, or NVIDIA Triton on k8s. For instance, TensorFlow Serving in docker provides an HTTP API, which allows clients to post image data and a prediction is returned as bounding box. In comparison, Triton Inference Server provides dynamic batching and multi-framework support on both CPU or GPU, which is useful for cloud inference requiring high performance and low latency.

6.9.2 Scaling and High Availability

The ability to scale elastically is a key benefit of cloud deployment. In cloud platforms such as Kubernetes(GKE, EKS), scaling is a few clicks away so long as your model-serving containers have auto-scaling functionality. As an illustration, if we deploy on Kubernetes using Helm charts, it is trivial to run multiple instances of your model ("pods"). The load balancer will take care of distributing inference requests and will automatically restart any containers that exit. Of course, services like Lightning AI (formerly LIT) now allow you to deploy Python-based image-processing APIs that are built with auto-scaling, making it even easier to handle variable demand.

6.9.3 Monitoring, Logging, and Model Management

Once your model is in production, it is very important to monitor the health of your model. Most cloud platforms have built-in features like drift detection (to inspect whether the distribution of incoming data is approaching the data distribution used to train the model), skew detection, performance logging (latency, error rate), and more. These features allow the user to set an alert or automate the retraining of the model when the systems detect drift that is beyond the specified bounds. Most Kubernetes and serverless deployments will also have some type of integration with the logging and monitoring services (CloudWatch, Stackdriver.).

6.9.4 Challenges in Cloud Deployment

1. Containerization Difficulties—Packaging models with all dependencies, installing GPU drivers, and configuring optimized run-times (Triton, TensorRT, or OpenVINO) poses great difficulties.
2. Resource Expenses—GPUs can be expensive, and static provisioning of GPUs could lead to sprawl and wasted resource. Auto-scaling helps, but tuning is important to the latency with splaying up pods and over-provisioning.
3. Latency vs. Throughput—The inference time is a function of batch size; as the batch size increase, utilization increases but this may lead to unacceptable latency. Dynamic frameworks like Triton, for example, can help to manage latency versus throughput compromises.
4. Model versioning and rollbacks—Deploying a new version of a model requires communication and coordination. Kubeflow and ModelOps pipelines may have supporting features for safe versioning, along with managing A/B test models and rollback strategies.
5. Data Privacy and Compliance—Sending images to the cloud may create data handling and privacy implications. There needs to be a plan for securing data while it is transit, secure the data for when it is at rest, and to have a plan for compliance/regulations under your local or regional guidelines, e.g., GDPR.

6.9.5 Real World Examples

- Google Cloud AI Platform: Train your object detection models using the TensorFlow Object Detection API, export them, containerize, and deploy on Cloud Run or AI Platform Prediction. You can configure skew and drift detection to monitor and assure that your model is reliable.
- AWS SageMaker and DeepLens: SageMaker can train and deploy models. You can serve them through managed endpoints or deploy them to an on-premises device, such as AWS DeepLens, for offline inference.
- NVIDIA Triton on Kubernetes: Inference frameworks for containers, such as Triton running on GKE or EKS, support multiple frameworks, host models concurrently, and enable dynamic batching; enabling maximum utilization of your GPU.

References

1. Yoon, H., Lee, S.H., Park, M.: TensorFlow with user friendly graphical framework for object detection API. arXiv preprint arXiv:2006.06385 (2020)
2. Toxigon: Object detection frameworks that will dominate 2025. Available online: https://toxigon.com/object-detection-frameworks-that-will-dominate-2025?utm_source=chatgpt.com (accessed on 26 September 2025)
3. Klingler, N.: PaddlePaddle: An open-source deep learning framework. viso.ai (26 March 2025). Available online: https://viso.ai/deep-learning/paddlepaddle/?utm_source=chatgpt.com (accessed on 26 September 2025)
4. OpenVINO. In: Wikipedia, the free encyclopedia (29 June 2025). Available online: https://en.wikipedia.org/wiki/OpenVINO?utm_source=chatgpt.com (accessed on 26 September 2025)
5. Oliveira, S.D., Topsakal, O., Toker, O.: Benchmarking automated machine learning (AutoML) frameworks for object detection. Information **15**, 63 (2024). https://doi.org/10.3390/info15010063
6. GeeksforGeeks: Introduction to TensorFlow (21 May 2025). Available online: https://www.geeksforgeeks.org/python/introduction-to-tensorflow/ (accessed on 26 September 2025)
7. Abadi, M., et al.: TensorFlow: A system for large-scale machine learning. arXiv preprint arXiv:1605.08695 (2016). Available online: https://arxiv.org/abs/1605.08695?utm_source=chatgpt.com (accessed on 26 September 2025)
8. Chollet, Francois. Deep learning with Python. simon and schuster, 2021
9. Imambi, Sagar, Kolla Bhanu Prakash, and G. R. Kanagachidambaresan. "PyTorch." Programming with TensorFlow: solution for edge computing applications. Cham: Springer International Publishing, 2021. 87–104
10. Kim, Seok Young, et al. "Extending the onnx runtime framework for the processing-in-memory execution." 2022 International Conference on Electronics, Information, and Communication (ICEIC). IEEE, 2022
11. Kim, Phil. "Matlab deep learning." With machine learning, neural networks and artificial intelligence 130.21 (2017): 151
12. Bradski, Gary, and Adrian Kaehler. Learning OpenCV: Computer vision with the OpenCV library. "O'Reilly Media, Inc.", 2008
13. Cengil, Emine, Ahmet Çınar, and Erdal Özbay. "Image classification with caffe deep learning framework." 2017 International Conference on Computer Science and Engineering (UBMK). IEEE, 2017
14. Li, Haidong, et al. "Research on overfitting of deep learning." 2019 15th international conference on computational intelligence and security (CIS). IEEE, 2019
15. Li, Zhuo, Hengyi Li, and Lin Meng. "Model compression for deep neural networks: A survey." Computers 12.3 (2023): 60
16. Goutam, Kelam, et al. "Layerout: Freezing layers in deep neural networks." SN Computer Science 1.5 (2020): 295

Chapter 7
Training and Fine-Tuning Object Detection Models

Abstract This chapter is dedicated to important concepts for training and fine-tuning object detection models. This chapter stresses the data preparation stage, which spans the real-world task of annotating and augmenting your data. This chapter also emphasizes the importance of transfer learning (especially for models that are limited in their training data) as there are many model architectures available, it is possible they could have been trained on a large dataset that is irrelevant to the task you are exploring. This chapter discusses the important hyperparameters such as learning rate and batch size, and how to automate their settings. The chapter explains the creation of multiple task loss functions, such as focal loss and Smooth L1 loss that need to be balanced between classification and localization. The chapter explains potential strategies for dealing with overfitting and underfitting—regularization, early stopping, and class imbalance. Overall, this chapter compiles a selection of best practices in the specificity of the geographical location to build a framework for model development that yields performance.

Training and fine-tuning object detection models involves data preparation, model architecture, loss function, and optimization strategies. Object detection models like YOLO, Faster R-CNN, and SSD use a two-stage process: feature extraction and bounding box prediction. In training, the model learns to classify objects and predict the bounding box with specific measurements. One of the key stages in training an object detection model is annotating the dataset, which entails labeling the objects with bounding boxes. While the annotation process can be tedious, the quality and volume of annotated data will impact how well the model performs. Data augmentation (e.g., random crop, flipping, color jitter) can increase data variability, thereby improving the model's robustness and generalization ability [1]. Transfer Learning makes a significant contribution towards fine-tuning object detection models. Almost always, you will need to use pre-trained weights (previous training on a large dataset e.g., ImageNet) that are used to initialize the model, rather than training a model from scratch. Training from pre-trained weights speeds up convergence in the model, and often results in superior models with improved accuracy, especially in cases of

A. Elhanashi and S. Saponara, *Deep Learning for Object Detection and Localization*,
https://doi.org/10.1007/978-981-95-4846-0_7

limited data [2]. Fine-tuning is training a model on a different, but related dataset by freezing certain layers in the model. This means only the last convolutional layer can be modified when fine-tuning the entire network, thus, only a small learning rate should be used to adjust the later layers for the target domain [3]. When training an object detection model, the choice of loss function is very important because depending on what loss function you use you could adjust the balance of the classification and the localization tasks. Most frameworks for object detection use a classification loss, such as cross-entropy, and a localization loss, such as Smooth L1 loss, which can fairly be called a regression loss function. Recently, focal loss [4], designed to combat class imbalance, has received a good deal of attention yet has not been used in a detection model. Focal loss applies a way to down-weight the easy negatives and quotes hard examples in its calculations, which allowed for significant improvements in accuracy of detection models trained on harder datasets [4]. Many of the optimization algorithms, such as limited Stochastic Gradient Descent (SGD) with momentum, are used for minimizing the combined loss function. Stochastic Gradient Descent (SGD) may also be adapted through various learning rate scheduling functions, such as step decay, which in some cases can formally schedule supported by work undertaken by [5], or even cosine annealing. The learning rate scheduling options can take a model getting stuck at local minima and another adjustments to improve the final performance obtained. In forming a model it is common practice to implement a stop early mechanism for the purpose of over fitting based on validation loss. Fine-tuning also involves hyperparameter tuning (batch size, number of epochs and data augmentation parameters). Small batch sizes can improve generalization, but can also lead to unstable gradients, while large batch sizes can encounter memory issues, but tend to provide smoother updates [6]. Grid search or Bayesian optimization methods can be used to search for optimal hyperparameters, which may result in better performance on a dataset and network architecture. New developments in fine-tuning methods include domain adaptation and few-shot learning; here, domain adaptation is tuned to minimize the domain gap between the pretrained source domain and the target domain through adversarial training or feature alignment to improve transferability [7]. Few-shot fine-tuning with meta-learning or prototype networks can adapt object detectors to novel classes with very few labeled examples [8].

7.1 Transfer Learning

Transfer learning has become a key strategy when training object detection models, especially in situations involving limited labeled data, or computational resources. Rather than training deep learning models from scratch, which requires vast amounts of annotated data, and the use of extensive computation, we instead utilize pretrained models (typically trained on extremely large scale datasets like ImageNet) when using transfer learning. This type of starting point opens up new tasks [9]. The use of pretrained models does not only speed up training, it often results in better performance, and improved generalisation in object detection models.

Central to the concept of transfer learning is the idea that, in CNNs, the early layers learn general features (edg, texture, simple shapes) that are essentially similar across different visual domains, and that the latter layers are particular learning details of the task. When we repurpose the pretrained weights of early layers, our object detection model can take advantage of the feature representations already learned, and do not need to (re)discover to treat the feature patterns of lowest level [10]. This is fundamentally important to the task of object detection, because in object detection the model must handle both classification and localization precisely.

In traditional transfer learning for object detection, the backbone network is initialized with pretrained weights (the feature extractor) and is then fine-tuned on the target dataset. Fine-tuning generally means retraining the latter layers in the architecture, and retraining the entire model or parts of the model with a smaller learning rate. This allows the model to adapt the higher-level representations to the specifics of the new dataset and retain the generalized features of the previous datasets [9, 10]. For example, in Faster R-CNN, most practitioners will use weights trained on ImageNet for the ResNet or VGG backbone, and then train the region proposal network (RPN) and detection heads from scratch or fine-tune them.

One key idea in transfer learning is determining which layers to freeze and which layers to part fine-tune. Although the first few layers freeze, the weights of the earlier layers will remain fixed and will not suffer from the generic feature representations lost on the new dataset when the target dataset is small or similar to the source dataset. However, fine-tuning more of the layers in the model may be a better strategy if the target dataset looks quite different to the source dataset mais also, it comes with an increased chance of overfitting to the new dataset [10]. There are several ideas to start fine-tuning the layers of a model with layer unfreezing starting with the top layers of the model and working backwards. This provides a level of control on the adaptations to the model [11].

Transfer learning also helps to alleviate overfitting, which is common when training deep networks with limited data. Since the pretrained model has learned general features, it will require less adjustment, which also mitigates the adversities of the model learning the noise or irrelevant properties of the smaller data population [9]. Additionally, transfer learning facilitates faster convergence, meaning the model will require fewer epochs to reach a satisfactory level of accuracy, thus reducing consumption of computational resources. While this has traditionally involved using a model pre-trained on ImageNet, more recent work uses large scale detection datasets, such as COCO or Open Images, where the model has greater context for the object localization task when compared to image classification. While the models can then be fine tuned or be used as varying detection problems, the models have retained the previous learning of object boundaries and shapes, if not also the prior learning also related to image classification [12]. In most cases, pre-trained on domain-specific data will offer a superior performance when compared to the pre-training on generic classification tasks, especially for detection tasks, thanks to domain-specific pre-training.

Domain adaptation is also a kind of transfer learning. Domain adaptation is concerned with the situation where the datasets between the source and target differ—either in visual appearance or in the data distribution causing deviations in the domain. An example might be where models pretrained on natural images can perform poorly when used in medical imaging or autonomous driving scenarios because of the domain shift. Now, the risk of domain shift applies only to target data collection. Thus, to enhance further the detection ability of the target domain, different strategies can be used during fine-tuning (e.g., adversarial training, feature alignment, and self-supervised learning). Domain adaptation could be portrayed as a transfer learning task that seeks to deal with differences of visual appearance between domains explicitly [13].

It can also be noted that few-shot learning approaches follow transfer learning principles more closely as the focus is on enabling these models to generalize from very limited labeled data when learning new classes or domains. Essentially, you want your model to learn how learn fast, and which is the goal of the meta-learning frameworks [14]. For example, few-shot fine-tuning for object detection can utilize pretrained detectors and make use of prototype learning or metric learning to generalize from a small number of annotated data, where few-shot learning would be useful when the cost of data collection is high, or data may not be captured in reality.

The process starts with a second-hand model (YOLOv8 or Faster R-CNN) initialized with COCO dataset weights that are imaginably strong about the task of object detection. The model is changed over to a custom dataset where images are labeled with ground truth bounding boxes (GT boxes) in green that signify the exact object locations and classes of interest. The custom dataset will be split into training, validation, and test sets for later evaluations, and essentially the model will be re-purposed. During re-purposing we will also modify the architecture to suit the new task. For YOLO, we will only change the detection head to directly predict the custom classes, while for Faster R-CNN it will also require the changes to the region proposal network (RPN) with a classifier. The custom dataset allows for models to be trained in two ways: fine-tune (updating all layers) or feature extraction (freeze the backbone and update the remaining layers). Data augmentation techniques (flipping, rotation, and scaling) can be applied in varying degrees in order for the model to discover better generalizations.

In the evaluation phase, model predictions (indicated with red boxes) are compared with ground truth (represented using green boxes). In this phase, evaluation metrics such as Intersection over Union (IoU) and mean Average Precision (mAP) can be useful. Further, it is a good idea to visualize predicted and GT boxes side-by-side to assess the localization accuracy of your model. The final stage is exporting your trained model to an optimized format such as ONNX or TensorRT so that it can be used on either an edge device or hosted on a cloud-based architecture. This workflow combines the powerful detection algorithms made available through FastAI and also the systematic process of training, validating, and testing to ensure high levels of accuracy when adapting these detection models to your own cases. The ASCII diagram struts these steps, and draws attention to key aspects like the comparison GT vs. predicted boxes and performance metrics.

7.2 Hyperparameter Optimization

Hyperparameter tuning is essential for enhancing the performance of object detection models. Hyperparameters differ from parameters in that parameters are learned during the training procedure, while hyperparameters are often configured before training and govern the learning method, learning speed, and final accuracy. Some of the most important hyperparameters in object detection are learning rate, batch size, number of epochs, optimization algorithm, weight decay, and hyperparameters which pertain to data augmentations and the model architecture, such as anchor boxes' width, height, and aspect ratios [15].

The learning rate is known to be the most important hyperparameter in the model. The learning rate hyperparameter effectively determines the step size in the weight updates during training. A learning rate that is too high will skip minima and fail to converge. On the other hand, a learning rate that is too small will slow down the training time and possibly end up in local minima [16]. Many object detection frameworks use learning rate schedules to train and configure the hyperparameters—step decay, cosine annealing, and cyclical learning rates are some examples—in an effective way in order to balance converging quickly while training later adjustments [17]. The batch size affects training dynamics and model generalization. Smaller batch sizes introduce noise in gradient estimation, which can serve as a sort of regularizer that improves generalization, but slower convergence due to less parallelism. Conversely, larger batch sizes lead to faster training but can cause the model to converge to sharp minima, resulting in poor generalisation [18]. Some have argued the learning rate should appear to scale with batch size to keep training dynamics consistent, a phenomenon called the linear scaling rule [19]. Also, the choice of optimizer: stochastic gradient descent (SGD), Adam, RMSProp or any of their many variants determines how the model explores the underlying loss landscape. SGD with momentum is still frequently used for object detection due to its perceived stability and good generalisation properties, however, adaptive optimizers such as Adam are on the rise particularly for faster convergence in the early phases of training. Adam has been viewed as having worse generalization compared to SGD [20], which has opened the door to hybrid methodologies and to change optimizers during training [21].

Weight decay (L2 regularization) is a widely used way to reduce overfitting by penalizing larger weights. The weight decay coefficient must be tuned because too much regularization will lead to underfitting, whereas too little will lead to overfitting [22]. Likewise, dropout rates, where used, can be tuned depending on the complexity of the dataset and its size. Anchor box parameters are hyperparameters that differ according to many object detectors—Faster R-CNN, SSD to name a few. By carefully right-sizing and aspect ratios of the boxes, anchor boxes provide the model with predicted bounding boxes that are similar to and located around the target boxes. Several AutoML techniques have been developed to optimize anchor configurations and demonstrate better detection performance [23]. The K-means heuristic approach (applied to ground-truth bounding boxes) is a popular choice for deciding anchor sizes that distribute around the dataset distribution [24]. Similarly, the data augmentation hyperparameters (rotation angle, scaling factor, probable flipping, and intensity of color jitter, etc.) improve model robustness, and reduce overfitting, which can impact detection effectiveness [25]. The strength of the augmentation must be chosen, and the augmentation type must generally be thought to avoid validation and test sample that too unrealistically interfere with the model. If human hyperparameter tuning is manual and is thus inefficient, it becomes even more difficult due to the number of hyperparameters influencing the complex relationships within deep learning models. Automated methods of hyperparameter optimization, such as grid search, random search, Bayesian optimization, and evolutionary algorithms, have gained popularity [26]. Random search outperformed grid search by exploring more diverse hyperparameter combinations in fewer trials [27]. Bayesian optimization explores the relationship between hyperparameters and model performance and uses this to model and explore the hyperparameter space to locate promising values [28]. Recently, hyperparameter tuning has been incorporated into neural architecture search (NAS) and multi-fidelity methods. Multi-fidelity methods, such as Hyperband, adaptively allocate computational resources (e.g. GPU cores) and discard poor Hyperparameter configurations early in training [29]. This selection is an important component of object detections which can be resource-intensive training processes. Hyperparameter optimization frameworks such as Optuna, Ray Tune, and Google Vizier provide scalable and reproducible hyperparameter tuning tools for deep learning models, including in object detection research and application in industry [30]. Their use will speed up model development cycles and provide more consistent performance improvements. Figure 7.1 shows the process of hyperparameter tuning. Different sets of hyperparameters (number of layers, number of neurons, learning

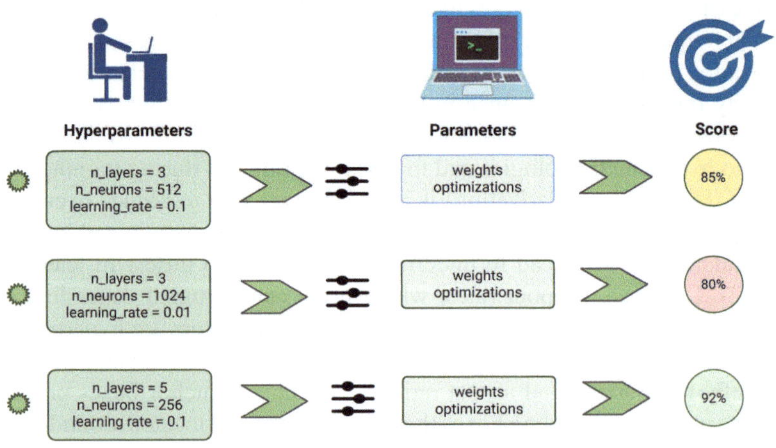

Fig. 7.1 The process of hyperparameter tuning. Different sets of hyperparameters (number of layers, number of neurons, learning rate) define distinct model architectures. Each model learns its own internal parameters (weights), resulting in different performance scores, allowing for the selection of the best configuration

rate) define distinct model architectures. Each model learns its own internal parameters (weights), resulting in different performance scores, allowing for the selection of the best configuration.

7.3 Loss Functions

Loss functions are essential for training object detection models as they describe how far the predicted outputs differ from the ground truth and lead to optimizations on the model parameters. Relative to classification tasks, object detection involves simultaneous optimization of two tasks: correctly classifying the object category and getting the bounding boxes direction correct. Thus, detection models typically use compound loss functions for their classification and localization loss.

In the classification task, the most widely used loss is cross-entropy loss or one of its many variants, e.g. focal loss. With cross-entropy loss, every example is weighed equally. However, in object detection tasks especially, the negative samples typically outnumber the positive samples. Thus, the majority of proposals or anchors will usually be background or negative samples and this imbalance has the potential to overwhelm the model with easy negative examples, which can substantially reduce the detection accuracy [31]. The focal loss by [32] specifically down-weights the loss assigned to well classified examples, and focuses training on hard examples that are down-weighted. This down-weighting of examples was particularly useful in dense detectors like RetinaNet, which increased performance on very difficult datasets. Smooth L1 loss (also called Huber loss) is often used for bounding box regression. Smooth L1 loss exhibits L2 loss behavior near zero but becomes linear for larger errors, which reduces the effect of outliers as compared to the L2 loss while still allowing for stable gradients [33]. This balance allows the model to regress bounding box coordinates without over-penalizing the model for large errors since they could disrupt the training process.

In some state-of-the-art detectors, other loss functions for localization have been proposed, which intended to improve the alignment between the predicted and ground truth boxes. Intersection over Union (IoU) loss tries to directly optimize the overlap between predicted and target bounding boxes, so predicted spatial accuracy is improved [34]. There are also extensions of IoU loss, such as Generalized IoU (GIoU) and Distance IoU (DIoU). GIoU and DIoU deal with some shortcomings of IoU, particularly when boxes do not overlap, to improve convergence and the final performance of localization [35].

Many object detection frameworks have a multi-task loss that combines classification and localization losses via a weighted sum, where you have to tune the weights to find the balance between these two tasks. Improperly weighting can cause the model to focus too heavily on either classification or localization, which results in poor detection [36]. For example, Faster R-CNN utilizes the multi-task loss that sums the classification cross-entropy loss and the smooth L1 localization loss, weighted equally for both. In contrast, RetinaNet utilizes focal loss in combination with smooth L1 localization loss.

Another important consideration for designing a loss function is to handle class imbalance, in addition to using focal loss. Online hard example mining (OHEM) involves dynamically selecting the most difficult samples when training, thereby reducing the easy negative affect on the loss, without actually changing the loss function [37]. OHEM works in combination with any loss function (e.g. focal loss) and studies had shown that it increases training accuracy and efficiency. Recent studies have examined loss functions optimized for particular detection problems such as small object detection, dense scenes, and multi-scale detection. For example, adaptive loss weighting could tune the weights at training time based on task difficulty or the relationships between the attributes of the sample, which provides a dynamical balance and improved generalization during detection tasks [38]. Furthermore, in some cases, IoU based metrics are included in the loss for classification as well to encourage both the classification confidence and localization quality are aligned [39].

7.4 Dealing with Overfitting and Underfitting

Overfitting and underfitting are very important issues when training object detection models. Overfitting occurs when a model remembers the training data, including every noise and detail that doesn't transfer to new data and underfitting refers to an insufficiently complicated model that fails to capture important characteristics of the training and test data since it has not been trained effectively or the model is too simple. To best manage both challenges requires good data handling and model architectures as well as training plans.

A basic procedure to handle overfitting is by utilizing data augmentation to increase the size and variability of your data artificially using image transformations such as image flipping, rotation, cropping and color jittering [40]. Data augmentation provides the model with variations of examples that increase the models ability to generalize. Again, make sure to not apply too many distortions and vast variations of augmentations or the model may train with distorted data.

Regularization strategies like weight decay (or L2 regularization); penalizes the increase in large parameters, encouraging simpler models have less chance of overfit. Dropout randomly drops neurons during training; thus, dropping feature co-adaptation, lessening model overfit and making the model more robust. Dropout's effectiveness is more limited for convolutional layers—added in many detection architectures.

Early stopping observes how validation performance changes over time and stops training when performance fails to improve performance on unshown validation data. This process limits overfit by enforcing that no more than so many epochs are trained. It is important that the validation set was sampled needs to be a strong representation of the intended deployment scenario.

Finding the right degree of model complexity is key: Models too complicated are prone to overfitting on a small dataset, models too simple are also prone to underfitting complex tasks. Model pruning and knowledge distillation are examples of methods that reduce model complexity while maintaining accuracy.

Transfer learning is now traditional practice to avoid overfitting, especially with limited labelled data available. Pre-trained backbones replicate generalized features, meaning less task evaluation happens in the new training dataset. There must still be careful management of fine-tuning plans, the learning schedules, and the freezing plans so that the model doesn't risk undergoing underfitting or catastrophic forgetting.

Batch normalization normalizes layer inputs and stabilizes training, which speeds up convergence and acts as a regularizer to reduce overfitting [47]. Similarly, using appropriate optimizers and learning rate schedules steers the model toward flatter loss minima, with the potential to improve model generalizability [48]. Reliable evaluation via cross-validation protects against overfitting by evaluating the model performance on different splits of the data, though this comes at a higher computational cost [49].

In the case of object detection, the task is particularly complex because of class imbalance, as negative (background) samples often greatly outnumber positive (object) samples. While there are loss-based methods such as focal loss [50] which reduce the relative loss contribution of easy negatives and shift training's focus toward hard examples, there are also complementary measures like online hard example mining (OHEM), which is a mechanism for adapting the training process and dynamically selecting the hard samples. Underfitting is the opposite problem, and is more simply organized into three broader areas: give it enough time, use the right sized model, and hyperparameters. Too little training time, or too aggressive regularization can easily result in too little opportunities to learn, or too restrictions on learning meaningful representations [51]. Table 7.1 illustrates a summary of common techniques for improving deep learning model generalization and preventing overfitting.

7.4.1 Best Practises

1. Utilize Data Augmentation to max potential but when using transformation, maintain realistic values with augmentations to augment variability in data but not confuse the model
2. Use Regularization Methods (weight decay, dropout) sensibly to minimize overfitting, particularly in small datasets
3. Use Early Stopping; use a valid dataset and stop training before the model memorizes noise
4. Use Model Complexity Wisely, required based on model size relative to dataset size and task, prune or distill when size needs to be reduced

Table 7.1 A summary of common techniques for improving deep learning model generalization and preventing overfitting

Technique	Purpose	Pros	Cons	References
Data augmentation	Increase data diversity	Improves generalization; inexpensive	May introduce unrealistic samples if overused	[40]
Weight decay	Regularization	Reduces overfitting by penalizing large weights	Requires tuning of decay coefficient	[41]
Dropout	Regularization	Prevents neuron co-adaptation; robust training	Less effective in conv layers	[42]
Early stopping	Prevent over-training	Simple; stops training at validation plateau	Needs representative validation set	[43]
Model complexity adjustment	Balance over/underfitting	Pruning/distillation reduces size without accuracy loss	Requires extra steps and tuning	[44, 45]
Transfer learning	Improve generalization	Reduces overfitting on small datasets	Risk of forgetting if not carefully tuned	[46]
Batch normalization	Stabilize training	Speeds up training; acts as regularizer	Adds computation overhead	[47]
Optimizer and LR schedule	Better convergence	Leads to flatter minima; improved generalization	Needs careful hyperparameter tuning	[48]
Cross-validation	Robust model evaluation	Detects overfitting early	Computationally expensive	[49]
Class imbalance handling (Focal Loss, OHEM)	Focus on hard samples	Improves learning on challenging examples	Adds complexity to training	[50]
Sufficient training and capacity	Avoid underfitting	Ensures meaningful pattern learning	Overtraining risk if unchecked	[51]

5. Use Transfer Learning; utilize learned general features and reduce overfitting by fine-tuning trained backbones
6. Use Batch Normalization; implicitly provides regularization and increases training stability
7. Use Optimizer and Learning Rate Schedules to promote convergence towards flatter minima to improve model generalization
8. Use focal loss or OHEM to Combat Class Imbalance; make losses focus points firsthand on the important hard samples

9. Use Cross-Validation when possible; to detect overfitting and where to with hyperparameter tuning with more clarity
10. Undertrain; once again make sure the model has adequate training time and capacity, but make sure not to run too long.

References

1. Shorten, C., Khoshgoftaar, T.M.: A survey on image data augmentation for deep learning. J. Big Data **6**(1), 60 (2019). https://doi.org/10.1186/s40537-019-0197-0
2. Huh, M., Agrawal, P., Efros, A.A.: What makes ImageNet good for transfer learning? arXiv preprint arXiv:1608.08614 (2016). https://arxiv.org/abs/1608.08614
3. Yosinski, J., Clune, J., Bengio, Y., Lipson, H.: How transferable are features in deep neural networks? In: Adv. Neural Inf. Process. Syst. (NeurIPS) 27, pp. 3320–3328 (2014)
4. Lin, T.Y., Goyal, P., Girshick, R., He, K., Dollár, P.: Focal loss for dense object detection. In: Proc. IEEE Int. Conf. Comput. Vis. (ICCV), pp. 2980–2988 (2017). https://doi.org/10.1109/ICCV.2017.324
5. Loshchilov, I., Hutter, F.: SGDR: Stochastic gradient descent with warm restarts. arXiv preprint arXiv:1608.03983 (2017). https://arxiv.org/abs/1608.03983
6. Masters, D., Luschi, C.: Revisiting small batch training for deep neural networks. arXiv preprint arXiv:1804.07612 (2018). https://arxiv.org/abs/1804.07612
7. Ganin, Y., Lempitsky, V.: Unsupervised domain adaptation by backpropagation. In: Proc. 32nd Int. Conf. Mach. Learn. (ICML), vol. 37, pp. 1180–1189 (2015)
8. Wang, Y., Yao, Q., Kwok, J.T., Ni, L.M.: Generalizing from a few examples: A survey on few-shot learning. ACM Comput. Surv. **53**(3), 63 (2020). https://doi.org/10.1145/3386252
9. Shorten, C., Khoshgoftaar, T.M.: A survey on image data augmentation for deep learning. J. Big Data **6**(1), 60 (2019). https://doi.org/10.1186/s40537-019-0197-0
10. Huh, M., Agrawal, P., Efros, A.A.: What makes ImageNet good for transfer learning? arXiv preprint arXiv:1608.08614 (2016). https://arxiv.org/abs/1608.08614
11. Yosinski, J., Clune, J., Bengio, Y., Lipson, H.: How transferable are features in deep neural networks? In: Adv. Neural Inf. Process. Syst. (NeurIPS) 27, pp. 3320–3328 (2014)
12. Shorten, C., Khoshgoftaar, T.M.: A survey on image data augmentation for deep learning. J. Big Data **6**(1), 60 (2019) [Duplicate entry; replace with correct reference if required]
13. Ganin, Y., Lempitsky, V.: Unsupervised domain adaptation by backpropagation. In: Proc. 32nd Int. Conf. Mach. Learn. (ICML), vol. 37, pp. 1180–1189 (2015)
14. Wang, Y., Yao, Q., Kwok, J.T., Ni, L.M.: Generalizing from a few examples: A survey on few-shot learning. ACM Comput. Surv. **53**(3), 63 (2020). https://doi.org/10.1145/3386252
15. Li, L., Jamieson, K., DeSalvo, G., Rostamizadeh, A., Talwalkar, A.: Hyperband: A novel bandit-based approach to hyperparameter optimization. J. Mach. Learn. Res. **18**(185), 1–52 (2017)
16. Goodfellow, I., Bengio, Y., Courville, A.: Deep Learning. MIT Press, Cambridge (2016)
17. Loshchilov, I., Hutter, F.: SGDR: Stochastic gradient descent with warm restarts. In: Int. Conf. Learn. Represent. (ICLR) (2017)
18. Keskar, N.S., Mudigere, D., Nocedal, J., Smelyanskiy, M., Tang, P.T.P.: On large-batch training for deep learning: Generalization gap and sharp minima. In: Int. Conf. Learn. Represent. (ICLR) (2017)
19. Goyal, P., Dollár, P., Girshick, R., Noordhuis, P., Wesolowski, L., Kyrola, A., et al.: Accurate, large minibatch SGD: Training ImageNet in 1 hour. arXiv preprint arXiv:1706.02677 (2017)
20. Kingma, D.P., Ba, J.: Adam: A method for stochastic optimization. In: Int. Conf. Learn. Represent. (ICLR) (2015)
21. Wilson, A.C., Roelofs, R., Stern, M., Srebro, N., Recht, B.: The marginal value of adaptive gradient methods in machine learning. In: Adv. Neural Inf. Process. Syst. (NeurIPS) 30 (2017)

22. Krogh, A., Hertz, J.A.: A simple weight decay can improve generalization. In: Adv. Neural Inf. Process. Syst. (NeurIPS) 4, pp. 950–957 (1992)
23. Ghiasi, G., Lin, T.Y., Le, Q.V.: NAS-FPN: Learning scalable feature pyramid architecture for object detection. In: Proc. IEEE Conf. Comput. Vis. Pattern Recognit. (CVPR), pp. 7036–7045 (2019)
24. Redmon, J., Divvala, S., Girshick, R., Farhadi, A.: You only look once: Unified, real-time object detection. In: Proc. IEEE Conf. Comput. Vis. Pattern Recognit. (CVPR), pp. 779–788 (2016)
25. Shorten, C., Khoshgoftaar, T.M.: A survey on image data augmentation for deep learning. J. Big Data **6**(1), 60 (2019) [Duplicate; see r208, r216, r219]
26. Bergstra, J., Bengio, Y.: Random search for hyper-parameter optimization. J. Mach. Learn. Res. **13**, 281–305 (2012)
27. Bergstra, J., Bardenet, R., Bengio, Y., Kégl, B.: Algorithms for hyper-parameter optimization. In: Adv. Neural Inf. Process. Syst. (NeurIPS) 24 (2011)
28. Snoek, J., Larochelle, H., Adams, R.P.: Practical Bayesian optimization of machine learning algorithms. In: Adv. Neural Inf. Process. Syst. (NeurIPS) 25, pp. 2951–2959 (2012)
29. Li, L., Jamieson, K., DeSalvo, G., Rostamizadeh, A., Talwalkar, A.: Hyperband: A novel bandit-based approach to hyperparameter optimization. J. Mach. Learn. Res. **18**(185), 1–52 (2017) [Duplicate; see r222]
30. Akiba, T., Sano, S., Yanase, T., Ohta, T., Koyama, M.: Optuna: A next-generation hyperparameter optimization framework. In: Proc. 25th ACM SIGKDD Int. Conf. Knowl. Discov. Data Min. (KDD), pp. 2623–2631 (2019)
31. Lin, T.Y., Maire, M., Belongie, S., Hays, J., Perona, P., Ramanan, D., et al.: Microsoft COCO: Common objects in context. In: Eur. Conf. Comput. Vis. (ECCV), pp. 740–755 (2014)
32. Lin, T.Y., Goyal, P., Girshick, R., He, K., Dollár, P.: Focal loss for dense object detection. In: Proc. IEEE Int. Conf. Comput. Vis. (ICCV), pp. 2980–2988 (2017)
33. Girshick, R.: Fast R-CNN. In: Proc. IEEE Int. Conf. Comput. Vis. (ICCV), pp. 1440–1448 (2015)
34. Yu, J., Wang, J., Peng, Y., Gao, C., Yu, G., Sang, N.: UnitBox: An advanced object detection network. In: Proc. IEEE Int. Conf. Comput. Vis. (ICCV), pp. 5160–5168 (2016)
35. Rezatofighi, H., Tsoi, N., Gwak, J., Sadeghian, A., Reid, I., Savarese, S.: Generalized intersection over union: A metric and a loss for bounding box regression. In: Proc. IEEE Conf. Comput. Vis. Pattern Recognit. (CVPR), pp. 658–666 (2019)
36. Ren, S., He, K., Girshick, R., Sun, J.: Faster R-CNN: Towards real-time object detection with region proposal networks. In: Adv. Neural Inf. Process. Syst. (NeurIPS) 28, pp. 91–99 (2015)
37. Shrivastava, A., Gupta, A., Girshick, R.: Training region-based object detectors with online hard example mining. In: Proc. IEEE Conf. Comput. Vis. Pattern Recognit. (CVPR), pp. 761–769 (2016)
38. Kendall, A., Gal, Y., Cipolla, R.: Multi-task learning using uncertainty to weigh losses for scene geometry and semantics. In: Proc. IEEE Conf. Comput. Vis. Pattern Recognit. (CVPR), pp. 7482–7491 (2018)
39. Pang, J., Chen, K., Shi, J., Feng, H., Ouyang, W., Lin, D.: Libra R-CNN: Towards balanced learning for object detection. In: Proc. IEEE Conf. Comput. Vis. Pattern Recognit. (CVPR), pp. 821–830 (2019)
40. Shorten, C., Khoshgoftaar, T.M.: A survey on image data augmentation for deep learning. J. Big Data **6**(1), 60 (2019)
41. Krogh, A., Hertz, J.A.: A simple weight decay can improve generalization. In: Adv. Neural Inf. Process. Syst. (NeurIPS) 4, pp. 950–957 (1992)
42. Srivastava, N., Hinton, G., Krizhevsky, A., Sutskever, I., Salakhutdinov, R.: Dropout: A simple way to prevent neural networks from overfitting. J. Mach. Learn. Res. **15**(1), 1929–1958 (2014)
43. Prechelt, L.: Early stopping—but when? In: Orr, G.B., Müller, K.R. (eds.) Neural Networks: Tricks of the Trade, pp. 55–69. Springer, Berlin (1998)
44. Zhang, C., Bengio, S., Hardt, M., Recht, B., Vinyals, O.: Understanding deep learning requires rethinking generalization. In: Int. Conf. Learn. Represent. (ICLR) (2017)

45. Han, S., Pool, J., Tran, J., Dally, W.J.: Learning both weights and connections for efficient neural networks. In: Adv. Neural Inf. Process. Syst. (NeurIPS) 28, pp. 1135–1143 (2015)
46. Huh, M., Agrawal, P., Efros, A.A.: What makes ImageNet good for transfer learning? arXiv preprint arXiv:1608.08614 (2016)
47. Ioffe, S., Szegedy, C.: Batch normalization: Accelerating deep network training by reducing internal covariate shift. In: Int. Conf. Mach. Learn. (ICML), pp. 448–456 (2015)
48. Keskar, N.S., Mudigere, D., Nocedal, J., Smelyanskiy, M., Tang, P.T.P.: On large-batch training for deep learning: Generalization gap and sharp minima. In: Int. Conf. Learn. Represent. (ICLR) (2017)
49. Kohavi, R.: A study of cross-validation and bootstrap for accuracy estimation and model selection. In: Proc. Int. Joint Conf. Artif. Intell. (IJCAI), pp. 1137–1145 (1995)
50. Lin, T.Y., Goyal, P., Girshick, R., He, K., Dollár, P.: Focal loss for dense object detection. In: Proc. IEEE Int. Conf. Comput. Vis. (ICCV), pp. 2980–2988 (2017)
51. Goodfellow, I., Bengio, Y., Courville, A.: Deep Learning. MIT Press, Cambridge (2016)

Chapter 8
Challenges of Object Detection and Localization

Abstract This chapter discusses the most important issues facing modern object detection and localization systems. It begins with object appearance variability, due to changes in lighting, occlusion, viewpoints, and intra-class variation, because reliable feature extraction is impossible under frequently encountered variance. The chapter likewise discusses the issue of scale variation, including how difficult it is to detect objects of different size, as well as the trade-offs obtained with image resolution. Speed-accuracy trade-off for real-time applications and computational bottleneck with edge deployment is further discussed. Finally, these discussions include class imbalance and background clutter, which result in bias in model training and false detections. While expounding these key issues, the chapter synthesizes these issues to give a comprehensive introduction to the research of better robust and more generalizable detection systems.

8.1 Variability in Object Appearance

Models for object detection and localization encounter enormous obstacles arising from object variability. As mentioned previously, lighting, occlusion perspective, and intra-class variation all lead to objects looking quite different despite possibly being the same object. For example, you could have an object under direct sunlight and same object under low light, which could account for performance differences in the model. Objects with partial occlusion also present challenges in object detection for algorithms that rely on an entire or relatively entire object structure to detect. The perspective a person sees an object from is also a challenge because a person could see an object in many different visual characteristics from different perspectives. A car when viewed from the front is going to be very differently shaped and thus have a significantly different texture than when viewed from the side or overhead. Finally, intra-class variation created by changes in color, size, and texture among different objects of the same class make it difficult for models trained on limited datasets. Chairs, for example, come in many different designs, however, a detection system must group them all into one class.

A. Elhanashi and S. Saponara, *Deep Learning for Object Detection and Localization*,
https://doi.org/10.1007/978-981-95-4846-0_8

Academic research also articulates these challenges consistently, with studies suggesturing that reliable feature extraction and data augmentation mechanisms are necessary for improved generalization. Proposed approaches, such as adversarial training, multi-view learning, and synthetic data generation, aim to account for problems related to variability, and yet achieving consistent performance due to variability in "real world" conditions has yet to solved.

8.1.1 Lighting and Illumination Changes

Lighting changes have pronounced effects on object detection performance. Depending on the light source or environment, the segmentation of objects will change—shadows, highlights, and low-light will hide essential object features, which can lead to misclassification or localization errors. For instance, [1] showed that convolutional neural networks (CNNs) trained on datasets with ideal lighting conditions perform poorly in low-light, unless those models have been explicitly trained on augmented datasets. As another example, [2] identified that extreme lighting conditions (e.g., glare, underexposure, etc.) can reduce the accuracy of detection by 40%, indicating the extreme sensitivity of deep learning models to changes in illumination.

There are different challenges related to lighting. With bright light conditions, overexposure can completely washout textures and edges, while shadows can create false contours that lead detection algorithms astray. Low-light conditions may also create noise, fog, or large portions of darkness which will not allow a model to find object boundaries. In the best conditions, histogram equalization would help with these more extreme environmental issues; however, better approaches are adaptive normalization or infrared-spectrum fusion which retains discriminatory features better than histogram equalization.

In addition, temporal lighting changes—such as shifts from day to night—are demanding models to generalize over dynamic conditions. Some methods circumvent this issue with multi-exposure training datasets or transfer learning domain adaptation techniques. But in real world scenarios (e.g., autonomous driving or surveillance) greater robustness is required, because inconsistent lighting remains an important failure mode of deployed systems. Future work may also develop self-supervised learning or use physics- based augmentations to help increase invariance to light sources rendering empirically irrelevant lighting artifacts.

8.1.2 Occlusions and Partial Visibility

In real-world scenes, objects are frequently occluded by other objects, resulting in limited visual input that makes detection difficult even for the best detection systems. [3] noted that occlusion is a significant issue, especially in dense scenes like pedestrian detection and autonomous driving, where objects are heavily overlapping;

it happens frequently. Their analysis found that even state-of-the-art detectors like Faster R-CNN and YOLO showed a 20–30% fall in accuracy when the level of occlusion was higher than 50%, largely due to the occlusion of relevant features.

The extent of the occlusion depends on the percentage, firstly, of the object which is visible and the complexity of the scene. For instance, in cities, a pedestrian that can only be seen in-part behind a parked vehicle would have no visible limbs and facial features, which makes recognition difficult. Similarly, in retail environments, individual products on shelves are often stacked and obscured so that it is difficult for any solution to keep up with the states of inventory items. To address these issues, there have been many approaches from researchers. Approaches like attention mechanisms [4] allow the model to concentrate on observable regions while inferring missing sub-parts, while other approaches such as part-based models break-down objects into observable segments, and by being careful improves the robustness of the solutions when only partials are observable.

While progress has been made in handling occlusions, the problem remains open—particularly in dynamic situations where objects can drift unpredictably. Future advances presumably will bring in temporal data from video streams or 3D spatial information for predicting the occluded areas. However, reliable performance across many complex real-world situations remains a major research problem in object detection problems.

8.1.3 Viewpoint and Pose Variations

Depending on the angle or pose we observe an object from, it can appear very differently, which causes a great deal of difficulty for object detection systems. A car displayed in the front view, side view, or top view can look quite different and not only can key features such as shape, texture, and structural contours of that object change, the modality as a whole changes as once you distinguish between features, now you must learn multiple representations for the same object class, which creates extra complexity for the model and potential confusion leading to misclassification. Su et al. [5] conducted an analysis of how changes to the viewport impact deep learning model performance and found that with a multi view dataset, the robustness of the model improved as it learned multiple viewpoints, however, these types of datasets can take a considerable amount of time to collect and are often not feasible in many real-life situations where collecting all the relevant different viewpoints are complex and impractical.

This issue is even more exaggerated when referring to deformable objects or articulated poses. To illustrate, there are a multitude of configurations in which a person can appear; they can be standing, sitting, or bending over, where in each corresponding configuration you would want your model to recognize the same underlying class, even though the visual differences are considerable. Similarly, animals or other flexible objects can also undergo non-rigid transformations that complicate detection. Existing strategies to address these obstacles include data augmentation techniques, which artificially increase the variability of available training data (synthetic viewpoint generation, and geometric transformations), and some researchers are combining 3D reconstructions or using multi-camera arrangements to obtain multiple views of objects, however these approaches can add significant computational overhead.

Regardless of these advances, effort toward achieving viewpoint invariance is still ongoing. Future work might include leveraging self-supervised learning to reduce dependency on labelled multi-view data, or including some form of spatial reasoning in the model to estimate 3D structure from a collection of 2D images. Ultimately, the balancing act between model generalization, and training efficiency, will remain a significant hurdle for object detection systems, particularly when considering variance of pose and viewpoint.

8.1.4 Intra-Class Variability and Deformations

There can be a great deal of variability in shape, size, texture, and structure in objects that belong to the same category, and this can be quite problematic for object detection applications. The "chair" category comprised of rigid office chairs, curved lounge chairs, and simple stools, each of which possess different visual features. Even animals of the same species vary in terms of pose, fur pattern, or dimensions. [6] studied the variability of intra-class concepts in fine-grained object detection and note that non-rigid surface shapes (e.g. clothing, soft toys, articulated creatures) as a greater challenge due to having a structure that is non-rigid but varies infinitely in shape and is literally understood as belonging to the same category.

Additionally, objects can also exhibit dynamic deformations (e.g. a waving flag, a moving animal, folded fabric, etc.) and the inherent issues are often further complicated in circumstances where objects are undergoing extreme dynamic deformations that are also non-rigid. In such cases, traditional object detectors that are designed using synthetical geometric representations of features undertake great effort to accommodate all the physical movement, and deformable objects can range anywhere from small and tiny details to taped and figured objects that are supposed to be understood as belonging to the same class object.

In response, more recent architectures such as deformable convolutional networks (DCNs) [7] have been developed and relate to such suggestions. Deformable networks capture increasingly detailed components or visual properties of objects within a region of interest that change shape based on the object's inherent distortion, allowing for improved object detection accuracy for non-rigid instances because DCNs can learn to "warp" convolution kernels based on the underlying structure of deformable objects. Although these advances, much like the intra-class variability problem remains a clear obstacle, especially as we work on applications where objects can be seen in a multitude of forms and fashions. Future work may employ hybrid techniques that bring together geometric priors with learned deformable features, and apply generative models to provide various intra-class examples during training. However, overall a major challenge in computer vision is to have strong generalization across all variability possible in a class.

8.2 Scale and Resolution Differences

One of the main problems with object detection systems is their ability to cope with objects that are observed at very different scales and resolutions in the same scene. This variation may come from several aspects, such as the object distance from the sensor, the sensor's resolution capabilities and environment context. In an urban surveillance situation, for example, the same vehicle may be observed as a high-resolution image when it is nearby but as a cluster of pixels when it is very distant. The typical convolutional networks' fixed receptive fields make it extremely challenging to support this type of variation; often, it is just not possible to maintain object detection accuracy at all scales. Small objects are particularly difficult, since the features of small objects typically wash away after one or more pooling operations; while large objects may lack sufficient contextual information to classify at the object level.

In addressing these problems, modern computer vision systems have found several unique ways to solve and alleviate the challenges. One of the most common is the use of Feature Pyramid Networks (FPN), which is especially useful because it merges the shallow high-resolution features with the deep semantically aware features for multi-scale detection in a single pass. Another common method is multi-scale training, whereby input images may be resized at random during training to force the model to learn scale-invariant features. Similarly, attention mechanisms can also be applied to the input features, where the features at different scales can be weighed and prioritized. Thus the attention mechanism allows one to determine the most relevant level for each detected object. Most detection frameworks also use anchor-based mechanisms, whereby anchors of unmodified boxes at a predefined width or height with a range of aspect ratios are allowed to find better-suited boxes for objects dealing with object at different scales.

There is also the relationship between resolution and detection performance to consider. Small object detection often relies on high-resolution imagery to maintain key details, but high-resolution also requires processing more pixels, which takes up more processing power and time. This creates a fundamental tension between the two in system design, that is important in real-time applications (such as autonomous driving) where accuracy and speed are both desired. Current academic research is still exploring dynamic resolutions and super-resolution preprocessing methods that may mitigate this issue. Despite advances over years of research methodology, achieving consistent performance for all potential scales is still unresolved, which continues to stimulate novel work in neural architecture and multi-scale processing frameworks. Research continues to explore biological vision principles and new sensing technologies that could provide more graceful solutions to this ongoing problem with computer vision systems.

8.2.1 The Challenge of Scale Variation

As an image might contain objects varying greatly in size, from minuscule background masses to giant foreground items, size variation poses a fundamental problem for detection systems. This scale difference poses a recognition problem because classical CNNs recognize images through fixed receptive field sizes, thus are fundamentally limited in confronting extremes of large object dimensions.

- Problems with Small Object Detection: The detection of very small objects such as pedestrians in aerial imagery, distant traffic signs, or tiny products on retail shelves is especially difficult due to pixel representation. According to [8], objects smaller than 32×32 pixels suffer a large recall drop compared with those objects larger than 32×32 pixels. Their experiments on the COCO dataset have shown that popular detectors, such as YOLOv3 and Faster R-CNN, are lost for objects occupying less than 1% of the image area. The major concern comes into play during feature extraction where CNNs, applying successive downsampling processes through either pooling or strided convolutions, gradually lose tiny detailed information of small objects and finally leave behind very little discriminative information useful for both accurate classification and localization.
  To this problem, researchers have considered multi-scale feature fusion (for instance, Feature Pyramid Networks) or techniques that keep high resolution. Some of these methods preprocess with super-resolution or have separate layers for tiny-object detection. However, detecting ultra-fine details while saving on computation still remains a grand challenge, especially in real-time scenarios such as drone surveillance or medical imaging.
- Large Object Contextualization Issues: On the other hand, when objects are too big to be considered large (for instance, in the street scenes-a bus, or in satellite images-a building, or an indoor scan-a piece of furniture), problems arise. When a big object spans for most of the image, it is crucial that the model also consider fine details (such as textures, edges) and the global structural context so that they do not arrive at wrong assumptions. As per [9], detectors without architectural adaptation may wrongly segment parts of a large object as separate instances (for example, treating windows appearing on the facade of the building as separate objects).
  Possible solutions include adaptive receptive fields (for example, dilated convolutions) and multi-scale feature aggregation as means of retaining local and global information. Other frameworks employ spatial attention mechanisms to allow dynamic focusing on the areas of the object that are appropriate and relevant. However, the fundamental issue of maximizing precision in the observation of large objects of interest, while attention is paid to performance across a wider range of scales remains an important research area, especially in domains like autonomous driving and aerial mapping where scale is often an inherent complexity of the environment.

8.2.2 Multi-scale Detection Approaches

Scale variability—where objects of the same class appear at substantially different scales—is one of the challenges in object detection. A scene could contain objects at different scales (e.g., distant pedestrians versus nearby vehicles) and so, detectors must generalize across scales. Researchers have proposed a number of solutions;

1. Feature Pyramid Networks (FPNs): FPNs [10] improve detection by developing a multi-scale feature hierarchy of low-resolution semantically rich features (large objects) and high-resolution features (small objects). In other words, FPNs combine features from different levels of a backbone network (e.g., ResNet) so detectors can detect objects at different scales. It has been reported that FPN-based models improve small object detection by over 20% compared to single-scale detectors. As a result, they are a standard component of modern architectures - such as Mask R-CNN and RetinaNet.
2. Multi-Scale Training and Testing (Image Pyramids): Another possible solution involves training and testing detectors on resized versions of input images known as image pyramids. [11] showed that exposing models to various resolutions of images during training simply improves the scale robustness. Simply scaling input images by a random amount, the network was forced to learn scale-invariant features. The downside is that it increases computational resources as it demands redundant feature extraction over multiple scales. Some frameworks reduce redundancy by adopting scale aware weight sharing, but since they are deployed in real time, they may still favor accuracy over speed since there is a cost-benefit tradeoff.
3. Attention Mechanisms for Scale Adaptation: The latest advances include the use of attention modules to identify and select the appropriate scale while identifying and detecting objects. Scale-Aware Feature Aggregation (SAFA), for [12] example, introduces the ability to learn how to prioritize the feature maps at each pyramid level to maximize accuracy when identifying an object based on the level of views available. These models leverage the discriminative scales of modeled features at a computational cost that is lower than traditional, more rigid scales. Other variants allow for a cross-scale feature fusion or another form of reference to refine detections by context, which also includes capabilities of every resolution.

8.2.3 Impact of Image Resolution on Detection

Image resolution has an important impact on the performance of object detection systems, representing a fundamental tradeoff between salient detail and processing efficiency.

1. High Resolution Images—Higher resolution images (e.g., 4K+) provide closer detail, facilitate better detection for small objects and subtle feature recognition. However, they require considerably more processing resources, meaning

the inference speed will be slower (greater time to produce an estimated object) and require more memory. Liu et al. [13] showed that although 4K resolution improves detection of small objects (e.g., distant pedestrians, charge circles or small traffic signs), they increased other challenges such as a higher level of noise and greater irrelevant background. Noise and irrelevant background adds to the risk of increased false-positives as the model or object detection models can mistake the fine texture or artefacts added, as object-likelihoods. Also, high resolution processing requires more elaborate preprocessing (e.g. patch processing and adaptive down-scaling) to run efficiently without sacrifice the accuracy.

2. Low-Resolution Images: Low-resolution images, while some computation burden the same and are faster, can be suitable for real time applications like video surveillance, or implementations at the edge of the network. But down-sampling too severely (e.g., below 512 × 512 pixels), and detection accuracy drop-off, due to the loss of key distinguishing features finds that small objects especially are lost, as they blend with background noise during excessive downscaling, resulting in miss detections. For example, a license plate or a small handled item could potentially be lost entirely with aggressive compression. Some strategies counteract this with super-resolution or fuse multiple scales, but this adds more overhead for detection.
3. Optimal Resolution Strategies: The optimal resolution will depend on the end application. For example, high resolution may be needed for medical purposes, while lower resolution is required when speed is essential, for situations like drone navigation. Recent methods have also investigated the possibility of adaptive resolution selection, where the system determines the input resolution dynamically based on object density or scene complexity. Nevertheless, the search for the optimal resolution has not been settled as a suitable approach, while situations with varied dimension objects need to be efficiently managed [14].

8.2.4 Case Study: Autonomous Driving and Aerial Imagery

- Autonomous Vehicles: Autonomous vehicles operate in dynamic environments where they must detect objects at very different scales (nearby pedestrians and far-away traffic signs are much different in scales and sizes), representing a significant challenge for perception systems, since traditional object detectors do not usually account for this. Caesar et al. [15] found that fixed-scale detectors generally discard small objects and return false positives while processing high-resolution data in urban environments. Their study showed that reliable performance can only be expected from systems with adaptive resolution scales, allowing the system to dynamically shift focus as size and surrounding context change. To address this problem, nearby autonomous systems tend to take a multi-scale approach to feature fusion, which allows of the benefit of high-resolution spatial features while capturing deeper semantic features from lower-resolution layers (such as FPNs).

Some methods also utilize attention mechanisms to highlight relevant scales as well as consider context-aware posterior processing to prune unlikely detections. However, challenges remain with real-time processing and specifically with fast-moving scenarios that require low latencies. Future research could benefit from developing lightweight neural architectures that are optimized for multi-scale inference, or hybrid models that fuse LiDAR and camera approaches to improve the robustness of detection with respect to distance.

- Satellite and Drone Imagery: Satellite and drone imagery presents distinct and difficult challenges in object detection due to extreme scales. The objects of interest (e.g., cars, ships, or structures) may only be a few pixels in high-resolution imagery. Aerial imagery is not the same as standard object detection where the objects are relatively large and defined. Models will need to detect small targets in sparse images in complex backgrounds. Xia et al. [16] examined this problem in aerial imagery and provided known generalizations and novel architectures with high-resolution sub-pixel detail on feature extraction to help conserve computation times. They showed that standard detectors, designed for street vision, performed poorly in aerial context because they are 'too insensitive' to small features. A common method is to employ a high-resolution backbone network along with 'context modules' that add multi-scale information without very much down-sampling. Preparation methods (e.g., super-resolution processing), using deformable convolutions, and anchor-free detectors also showed promises for small object detection. Even with these, there are still issues with context conditions, poor visibility (e.g., clouds and shadows), and detecting objects with low contrast with the background. If the directions for the future of object detection on satellite imagery relates to the past, more self-supervised learning methods (e.g., less reliance on labeled datasets) and temporal context usage (e.g., on video-based satellite).

8.2.5 Future Directions and Solutions

Advancements in deep learning are creating fresh ways to face up to the scale variation problem in object detection. For example, Neural Architecture Search (NAS) for Scale Optimization used NAS to discover network architectures that are optimized for multi-scale object recognition [17]. While specific object detectors are designed or structured by hand, as those put together from various layers or boxes, the architectures generated by NAS are able to distribute its computational juice to the various scales adaptively, while also improving accuracy. This is an advantage in real-world applications where objects are very far away (pedestrian) and very close (car). Another new area with potential is the area of Hybrid Models that Combine CNNs with Transformers, which use both model architectures [18]. CNNs excel in extracting local features, while transformers provide, effectively, long-range context, which improves the contextual understanding of objects in the full range of scales. For example, a hybrid model could use a CNN to detect these fine grained visual details (small objects), while employing a transformer to maintain spatial coherence across the whole scene. This would act to diminish, at least in part, the performance drop we see by object scale, particularly in contexts such as aerial imagery or in chaotic urban environments, when the number of objects increases. In addition, Super Resolution for Small Objects using generative adversarial networks (GANs) has become an effective preprocessing step [19]. This technique of super-resolving small or blurry objects that are prior to detection helps to restore lost details that would otherwise impede the ability to be recognized. For instance, in satellite imagery, the super-resolution of small structures like cars or road markings would make it possible for them to be detected by standard models. However, there was a fine line that needed to be considered to avoid introducing artifacts that ultimately guided the detector in the wrong direction.

Collectively, these patterns and techniques reflect the beginnings of a shift towards greater adaptable and robust scale-invariant detection models. Nonetheless, aspects such as computational burden, and generalization through varying datasets and industries are still areas requiring further research.

8.3 Real-Time Processing Constraints

Computer vision models for object detection and localization are becoming integrated into real-world applications quickly, including self-driving vehicles, surveillance, and robotics, where speed of inference becomes just as important as inference accuracy. Inference accuracy can often be achieved sometimes with an increase in latencies, and it continues to be challenging to perform powerful and precise inference at low latency due to the intrinsic trade-off between increasing model capacity and increasing inference time. One example of deep learning architectures where inference latency is prioritized is single-shot detectors (SSD), while two-stage networks like Faster R-CNN can achieve more accurate results but require significantly more computational resources. In terms of computer vision models on edge devices, there are additional practical considerations as edge devices might not have the power of powerful high-end GPUs. Current trends in computer vision and on-device inference use model optimizations including model pruning, quantization, and knowledge distillation to reduce model scale while still providing similar accuracies, and emerging ideas such as neural architecture search (NAS) and hardware aware model design illustrate the potential for future advancements that might be able to optimize the inference capacity even further in an automated way.

Even with the knowledge that nice improvements have been made already, more work is needed in the area of inference under real-world conditions, as the challenges presented by varying scene complexity and the presence of dynamic environments can highly affect performance. It is quite possible that future research will look at object detection inference as an area where adaptability for different scenarios is implemented in a manner that optimizes the inference speed for each application and system, providing a reliable inference in varying and sometimes unpredictable scenarios.

8.3.1 The Speed-Accuracy Trade-Off

Real-time object detection requires the careful optimization of the underlying trade-off between accurate detection and efficient computation. Newer detection approaches can be grouped broadly into two different model families with their own performance characteristics:

- Two-Stage Detectors: Two-stage detectors (like Faster R-CNN [20]) use a region proposal network (RPN) to generate region proposals followed by region-of-interest (RoI) classification and bounding box regression. While detection accuracy is especially high (55.8% AP on COCO [22]), the complex set of steps is computationally intensive, limiting the maximum frame rate to 5–15 FPS and has tight restrictions, especially for applications that have latency sensitivity.
- One-Stage Model Detectors: One-stage model detectors (like YOLO [14] and SSD [13]) use a prediction architecture which has both bounding box predictions and class predictions feed from feature maps, which corresponds directly to "real-time" detection (>30 FPS). YOLOv4 exemplifies this with detection of 43.5% AP [22] and execution speed of 65 FPS. Additionally studies have indicated their relative weaknesses in small object detection and being able to deal with occlusion situations [21]. The performance gap separating these paradigms has been minimized with the help of several innovations:

1. Feature pyramid networks (FPNs), introduced for multi-scale detection
2. Amended loss functions (e.g. focal loss)
3. Optimizations to the architecture (i.e. depthwise separable convolution)
4. New research that focuses on application-specific optimizations where:
5. Hybrid approaches are often adopted for safety-critical systems (e.g. autonomous vehicles)
6. Edge devices have YOLO models in a quantized network variant
7. Two-stage approaches are still preferred for high-precision applications

Most recent benchmarks [22] indicate that one-stage detectors have the advantage in real-time applications (e.g. drone navigation and video analytics), but two-stage detectors are still the approach of choice for maximum spatial accuracy if the constraints of the computational architecture are relaxed from the one-stage nature. Future work may look to develop dynamic architectures that dynamically accommodate the relative ease or difficulty of the scene [21]. This may help to close the fundamental tradeoff between the two paradigms.

8.3.2 Hardware and Computational Bottlenecks

The ability to deploy object detection solutions on edge devices, such as embedded systems, drones, and smartphones, presents significant hardware constraints on performance and efficiency. While cloud-based architecture is not limited in computational resources or processing capabilities, deploying on edge devices creates a necessity for a balance between accuracy and efficiency using distinct hardware limitations in memory, power consumes, and processing capabilities.

- Limitations in Memory and Power: While high-performance GPUs (e.g., NVIDIA Tesla, Jetson) can accelerate deep learning inference significantly, the power consumption limits their practical use in battery-operated devices [23]. On the other hand, mobile processors (e.g., Qualcomm Snapdragon, Apple Neural Engine) allow for engagement in energy efficient architectures, operating with significant constraints in deploying large and complex models. This introduces a design choice where a developer must use efficient architectures in a comprehensive design approach using complementary methods. Some methods include distilling the teacher model into multiple smaller "student" networks that will operate using less compute resources.
- Latency in Distributed Systems: Cloud-based offloading can help alleviate some of the local processing limits, but delays in the network introduce uncertainty to the applicability of applications, like self-driving cars or drones where real-time performance is critical [24]. Although latency is reduced with edge-computing, where inference occurs locally, we must employ very aggressive model compression [15]. There are some solutions to compress models through pruning (removing lower importance neurons), quantization (lowering numerical precision), and knowledge distillation which can reduce model size with minimal loss in performance. While this makes the models smaller, it creates an open question in how to remain robust with those optimizations, especially in safety-critical situations. Future innovations will push towards hardware-aware neural architecture search (NAS) to automate the design of efficient models for certain edge devices, and hybrid approaches to model development that can dynamically leverage a selection of cloud and/or edge processing based on real-time characteristics.

8.3.3 Optimization Techniques for Real-Time Detection

To reach true real-time object detection we must address the balance between the walls of speed and accuracy through specialized optimization methods. These optimization methods will attempt to address computational bottlenecks from many different aspects (e.g., algorithmic, software-related, hardware-aware) with the purpose to enable deployment in latency-yielding and latency-sensitive locations (e.g., self-driving cars, robots, mobile vision).

(a) **Model Compression**

Model compression aims to achieve reduced computation overhead from detected quality:

- Quantization (or relegating data sizes to lower precision)—Reducing computation memory bandwidth from 32-bit floating-point (FP32) to 8-bit integers (INT8) will not only accelerate computational operations, it can provide significant size reduction. For instance, the use of INT8 quantization can have a 2–4× speedup in inference while having no significant performance drop on well-calibrated models (<1% drop) [25]. Commonly used methods for INT8 quantization are post-training quantization (PTQ) and quantization-aware training (QAT). Although both of these methods aim for INT8 quantization, PTQ is the more straightforward method, while QAT has been shown to retain accuracies better during model advancement when reaching limited-bit precision.
- Pruning (or redundancy removal)—Redundant filters or neurons can be removed to shorten model size and reduce FLOPS (e.g., via magnitude-based, lottery ticket, pruning, etc.). For instance, Tiny-YOLO is a pruned version of YOLOv3 that is 5× smaller and >10% less accurate [26]. Pruning from an iterative process during training will help to recover the accuracy in the sparsified model.

(b) **Architectures with Larger Inputs**

Fast architectures also take advantage of architectural innovation for speed:

- MobileNet and EfficientDet: These are architectures that use depthwise separable convolutions instead of regular convolutions and use these depthwise and pointwise operations separately in a chain so that the parameters can be reduced by 80–90% [27]. Also, EfficientDet is using a search called neural architecture search (NAS) to scale these speeds further.
- Transformer-Based Architectures: Even if Vision Transformers (ViTs) are accurate, their overhead and quadratic complexity limit the efficiency of inference and speedup of computations available. MobileViT and EdgeViT inform users about hybrid CNN-Transformer architectures or tokenize certain tokens in the input, while achieving real-time speeds on a mobile CPU [28].

(c) **Co-Design of Hardware and Software**

Optimized deployments take advantage of hardware-level optimizations:

- Framework/Library Integration: Frameworks such as TensorRT and ONNX Runtime will provide optimizations at the graph-optimization stage (layer fusions, kernel auto-tuning) for NVIDIA GPUs or CPUs, providing users with 2–5× speedup [29].

- Dedicated Accelerators: Custom integrated circuits also referred to as chips, such as Google cloud TPUs or Intel Movidius at the edge exploiting fixed-point math and higher parallelism of computations at low cost. FPGAs/ASICs (such as Tesla's Dojo, AMD/Xilinx FPGAs) are 10–100× faster than standard models by customizing compute strategies specifically for detection pipelines [30].

8.4 Class Imbalance and Background Clutter

Object detection systems contend with serious issues of class imbalance and background clutter, which can negatively influence model performance class imbalance is when some object classes occupy more of the training set than others. When object classes are not evenly represented in the training data, the model may perform well on common classes and poorly on the rare ones; this can be problematic in some situations when objects are infrequent, such as with important medical imaging or rare species detection. Background clutter further complicates detection, which includes additional false signals, in the form of texture, or other visuals that are similar enough to the target object that they are sometimes confused with the "real."

8.4.1 The Class Imbalance Problem

Object detection has an inherent issue of class imbalance which occurs when a small number of object classes dominate the training data, thus creating biased classifiers that discount poor performance on rare objects. The long-tail distribution of classes in the LVIS dataset [31] shows this, where one class had over 1,000× more examples than another class. This means that the models had >60% average precision (AP) with classes that occurred frequently, but had a <10% average precision on rare classes.

Class imbalance takes on 2 prominent forms:

1. Foreground-Background Imbalance: typically represented by a ~100:1 (background:object) ratio of pixels [32], resulting in detectors that suppress backgrounds regardless of whether or not an object appears.
2. Foreground Class Imbalance: when certain object classes consistently appear in the training set, resulting in poorer generalization to underrepresented minority classes.

Recently, advances have introduced a number of viable approaches:

- Repeat Factor Sampling (RFS) [33]: Method that dynamically oversamples rare class instances during training, producing a balanced representation of all parts.

- Equalized Focal Loss [34]: Extends standard focal loss by including a class-aware modulator to deal with extreme imbalance.
- Decoupled Training [35]: Decouples feature learning from classifier optimisation, allowing specific handling of tail classes.

While these methods are useful, they still cannot guarantee the maintenance of detection quality on all classes, and overfitting to rare examples is still a threat. Future work could address the issue of leveraging self-supervised pretraining or using generative data to augment the training sets to further minimize the impacts of imbalance.

8.4.2 Background Clutter Challenges

Complex backgrounds are problematic for object detection systems, and the issues commonly fall in two ways. First, there are a high number of false positives when the detector thinks the background pattern is a foreground object. The analysis of the COCO dataset illustrates that more than 35% of the errors in detection are from background confusion [36]. Second, the detector will be unable to determine true features because the background images contaminate the features, and this occurs most acutely for small objects where there is less pixel information [37]. This challenge is most clearly illustrated in natural backgrounds environments where the object blends into the scene, for example, animals in foliage, and vehicles in urban environments. There are multiple approaches to tackle background interference; however, current methods utilizes a combination of techniques. Attention mechanisms, notably Convolutional Block Attention Modules (CBAM) [38], help the model attend to the part of the object, while down weight features that confuse the object in the foreground from the background through spatial attention, and channel attention. Adversarial Training similarly applies a concept from Generatively Adversarial Networks (GANs) [39] to train models for features that are more discriminatory from cluttered backgrounds. Lastly, there is multi-task learning [40], to improve detection by predicting locations of objects and segmenting the background, forcing the model to understand contextual relationships better.

Nonetheless, background clutter is still a significant issue, especially in unconstrained environments. Future efforts might use more sophisticated context modeling techniques or hybrid approaches which use local features and global scene knowledge. Creating more extensive datasets that includes extreme variations in background could also help improve model generalization against this essential detection challenge in future.

8.4.3 Integrated Solutions

Numerous modern object detection systems are based on integrated paradigms that bring together various multiple complementary techniques to navigate challenging conditions including detection accuracy, class imbalance, and feature representation. Now, state-of-the-art frameworks use systematic optimization among various components of the detection pipeline to produce a more consistent, robust performance bazel across different scenarios.

As a prime example, Libra R-CNN [41] adopts a balanced sampling paradigm across three major domains of detection. A balanced approach is first applied, through IoU-balanced sampling and during the input stage for biased hard example mining. Second, a balanced pyramid based network ensures consistent feature extraction through balanced information flow among different levels of features. Third, it employs a balanced L1 loss function that modulates gradient contributions and helps prevent easy negatives from unduly influencing training. Balancing across these three levels provides a more consistent training experience and improved detection performance, especially on hard cases with diverse scales or appearances of object classes.

One more innovative solution is Seesaw Loss [42], which is designed to address the stubbornly hard problem of class imbalance. In contrast to fixed class weighting as seen in the previous examples, Seesaw Loss applies a penalty for misclassification that can change depending on a dataset's frequency for each class, as well as the relationships between similar classes. For classes that appear often enough, it can reduce penalties for misclassifying these classes to prevent the model from growing biased in the future, while classes that are rarely appearing, but misclassified regularly, would receive a larger penalty to help the model's learning. Furthermore, Seesaw Loss uses a compensation mechanism to maintain discrimination across easily-conflated categories. It is particularly useful for long-tailed datasets in which certain object categories appear well less than their counterparts.

These different types of integrated solutions represent the difference from incremental changes to system-level design for object detection. By working across many phases of the detection pipeline, from sampling to loss calculation, the ensemble or composite models achieve higher performance than traditional models. In the future this may lead to more tightly integrated systems or architectures that are automatically adjusted by meta-learning or other adaptive architectures based on the input characteristics of detections. These integrated, system-level approaches are likely to emerge as the norm well into the future, going forward into the highly-complex, real-world problems associated with object detection.

8.4.4 *Evaluation Metrics for Imbalanced Data*

Conventional evaluation measures, such as mean Average Precision (mAP), can yield misleading results for imbalanced datasets, since models that perform very well for a class with a frequent amount of instances may do poorly for a class with a rare amount of instances (and vice-versa). As a response to this limitation in conventional metrics for imbalanced datasets, several researchers have proposed custom metrics which allow for a richer understanding of model performance across all classes.

AP@Few evaluates the model's ability to detect the least frequently occurring 20% of classes [43], and helps reveal if the system misses rare objects, even though it may have high overall mAP. This is especially relevant for applications that involve rare detection of instances, such as surveillance, medical imaging, and other applications relating to endangered species.

AP@BG provides a way to evaluate the false positive rate for background regions [44], and can help assess whether the model is significantly over predicting objects in empty places (a common problem when negative examples dominate, causing a lack of background). This metric could also be considered for low false positive rates for model interpretability—for example, in contexts where false alarms for "empty" scenes can be concerning, such as self-driving vehicles or security/surveillance footage.

These focused metrics present a way to manage a model's performance and improve its performance, especially at the class imbalanced state, when additional measures (e.g., mAP) may be insufficient. Future work should be directed towards an adaptive way to build a threshold for these measures, or possibly focused-cost sensitive metrics.

References

1. Geiger, A., Lenz, P., Urtasun, R.: Are we ready for autonomous driving? The KITTI vision benchmark suite. In: Proc. IEEE Conf. Comput. Vis. Pattern Recognit. (CVPR), pp. 3354–3361 (2012)
2. Dai, J., Li, Y., He, K., Sun, J.: R-FCN: Object detection via region-based fully convolutional networks. In: Adv. Neural Inf. Process. Syst. (NeurIPS), pp. 379–387 (2016)
3. Ren, S., He, K., Girshick, R., Sun, J.: Faster R-CNN: Towards real-time object detection with region proposal networks. IEEE Trans. Pattern Anàl. Mach. Intell. **39**(6), 1137–1149 (2015)
4. Wang, X., Girshick, R., Gupta, A., He, K.: Non-local neural networks. In: Proc. IEEE Conf. Comput. Vis. Pattern Recognit. (CVPR), pp. 7794–7803 (2018)
5. Su, H., Maji, S., Kalogerakis, E., Learned-Miller, E.: Multi-view convolutional neural networks for 3D shape recognition. In: Proc. IEEE Int. Conf. Comput. Vis. (ICCV), pp. 945–953 (2015)
6. Lin, T.Y., RoyChowdhury, A., Maji, S.: Bilinear CNN models for fine-grained visual recognition. In: Proc. IEEE Int. Conf. Comput. Vis. (ICCV), pp. 1449–1457 (2015)
7. Zhu, X., Hu, H., Lin, S., Dai, J.: Deformable ConvNets v2: More deformable, better results. In: Proc. IEEE Conf. Comput. Vis. Pattern Recognit. (CVPR), pp. 9308–9316 (2019)
8. Lin, T.Y., Maire, M., Belongie, S., Hays, J., Perona, P., Ramanan, D., Dollár, P., Zitnick, C.L.: Microsoft COCO: Common objects in context. In: Eur. Conf. Comput. Vis. (ECCV), pp. 740–755 (2014)

9. He, K., Zhang, X., Ren, S., Sun, J.: Deep residual learning for image recognition. In: Proc. IEEE Conf. Comput. Vis. Pattern Recognit. (CVPR), pp. 770–778 (2016)
10. Lin, T.Y., Dollár, P., Girshick, R., He, K., Hariharan, B., Belongie, S.: Feature pyramid networks for object detection. In: Proc. IEEE Conf. Comput. Vis. Pattern Recognit. (CVPR), pp. 2117–2125 (2017)
11. Singh, B., Davis, L.S.: An analysis of scale invariance in object detection—SNIP. In: Proc. IEEE Conf. Comput. Vis. Pattern Recognit. (CVPR), pp. 3578–3587 (2018)
12. Chen, K., Pang, J., Wang, J., Xiong, Y., Li, X., Sun, S., et al.: Hybrid task cascade for instance segmentation. In: Proc. IEEE Conf. Comput. Vis. Pattern Recognit. (CVPR), pp. 4974–4983 (2019)
13. Liu, W., Anguelov, D., Erhan, D., Szegedy, C., Reed, S., Fu, C.Y., Berg, A.C.: SSD: Single shot multibox detector. In: Eur. Conf. Comput. Vis. (ECCV), pp. 21–37 (2016)
14. Redmon, J., Farhadi, A.: YOLOv3: An incremental improvement. arXiv preprint arXiv:1804.02767 (2018)
15. Caesar, H., Bankiti, V., Lang, A.H., Vora, S., Liong, V.E., Xu, Q., et al.: nuScenes: A multimodal dataset for autonomous driving. In: Proc. IEEE/CVF Conf. Comput. Vis. Pattern Recognit. (CVPR), pp. 11621–11631 (2020)
16. Xia, G.S., Bai, X., Ding, J., Zhu, Z., Belongie, S., Luo, J., et al.: DOTA: A large-scale dataset for object detection in aerial images. In: Proc. IEEE Conf. Comput. Vis. Pattern Recognit. (CVPR), pp. 3974–3983 (2018)
17. Zoph, B., Le, Q.V.: Neural architecture search with reinforcement learning. In: Int. Conf. Learn. Represent. (ICLR) (2017)
18. Dosovitskiy, A., Beyer, L., Kolesnikov, A., Weissenborn, D., Zhai, X., Unterthiner, T., et al.: An image is worth 16x16 words: Transformers for image recognition at scale. In: Int. Conf. Learn. Represent. (ICLR) (2021)
19. Wang, X., Yu, K., Wu, S., Gu, J., Liu, Y., Dong, C., Qiao, Y., Change Loy, C.: ESRGAN: Enhanced super-resolution generative adversarial networks. In: Eur. Conf. Comput. Vis. Workshops (ECCVW) (2018)
20. Ren, S., He, K., Girshick, R., Sun, J.: Faster R-CNN: Towards real-time object detection with region proposal networks. IEEE Trans. Pattern Anal. Mach. Intell. **39**(6), 1137–1149 (2015)
21. Huang, J., Rathod, V., Sun, C., Zhu, M., Korattikara, A., Fathi, A., et al.: Speed/accuracy trade-offs for modern convolutional object detectors. In: Proc. IEEE Conf. Comput. Vis. Pattern Recognit. (CVPR), pp. 7310–7311 (2017)
22. Bochkovskiy, A., Wang, C.Y., Liao, H.Y.M.: YOLOv4: Optimal speed and accuracy of object detection. arXiv preprint arXiv:2004.10934 (2020)
23. Howard, A., Sandler, M., Chu, G., Chen, L.C., Chen, B., Tan, M., et al.: Searching for MobileNetV3. In: Proc. IEEE/CVF Int. Conf. Comput. Vis. (ICCV), pp. 1314–1324 (2019)
24. Chen, T., Moreau, T., Jiang, Z., Zheng, L., Yan, E., Shen, H., et al.: TVM: An automated end-to-end optimizing compiler for deep learning. In: USENIX Symp. Oper. Syst. Des. Implement. (OSDI), pp. 578–594 (2018)
25. Jacob, B., Kligys, S., Chen, B., Zhu, M., Tang, M., Howard, A., et al.: Quantization and training of neural networks for efficient integer-arithmetic-only inference. In: Proc. IEEE Conf. Comput. Vis. Pattern Recognit. (CVPR), pp. 2704–2713 (2018)
26. Molchanov, P., Tyree, S., Karras, T., Aila, T., Kautz, J.: Pruning convolutional neural networks for resource efficient inference. In: Int. Conf. Learn. Represent. (ICLR) (2017)
27. Tan, M., Pang, R., Le, Q.V.: EfficientDet: Scalable and efficient object detection. In: Proc. IEEE/CVF Conf. Comput. Vis. Pattern Recognit. (CVPR), pp. 10781–10790 (2020)
28. Mehta, S., Rastegari, M.: MobileViT: Light-weight, general-purpose, and mobile-friendly vision transformer. In: Int. Conf. Learn. Represent. (ICLR) (2022)
29. NVIDIA: TensorRT: High-performance deep learning inference. NVIDIA Developer Documentation (2021)
30. Jouppi, N.P., Young, C., Patil, N., Patterson, D., Agrawal, G., Bajwa, R., et al.: In-datacenter performance analysis of a tensor processing unit. In: ACM/IEEE Int. Symp. Comput. Archit. (ISCA), pp. 1–12 (2017)

31. Gupta, A., Dollar, P., Girshick, R.: LVIS: A dataset for large vocabulary instance segmentation. In: Proc. IEEE Conf. Comput. Vis. Pattern Recognit. (CVPR), pp. 5356–5364 (2019)
32. Lin, T.Y., et al.: Focal loss for dense object detection. In: Proc. IEEE Int. Conf. Comput. Vis. (ICCV), pp. 2980–2988 (2017)
33. Tan, J., et al.: Equalization loss for long-tailed object recognition. In: Proc. IEEE Conf. Comput. Vis. Pattern Recognit. (CVPR), pp. 11662–11671 (2020)
34. Zhang, S., et al.: Equalized focal loss for dense long-tailed object detection. In: Proc. IEEE Conf. Comput. Vis. Pattern Recognit. (CVPR), pp. 6990–6999 (2021)
35. Kang, B., et al.: Decoupling representation and classifier for long-tailed recognition. In: Int. Conf. Learn. Represent. (ICLR) (2019)
36. Zoph, B., et al.: Rethinking pre-training and self-training. In: Adv. Neural Inf. Process. Syst. (NeurIPS) 33 (2020)
37. Wang, T., et al.: Noise-aware fully webly supervised object detection. In: Proc. IEEE Conf. Comput. Vis. Pattern Recognit. (CVPR), pp. 11326–11335 (2020)
38. Woo, S., et al.: CBAM: Convolutional block attention module. In: Eur. Conf. Comput. Vis. (ECCV), pp. 3–19 (2018)
39. Wang, X., et al.: VAT: A self-supervised technique for object detection. In: Proc. IEEE Int. Conf. Comput. Vis. (ICCV), pp. 2385–2394 (2019)
40. Misra, I., et al.: Cross-stitch networks for multi-task learning. In: Proc. IEEE Conf. Comput. Vis. Pattern Recognit. (CVPR), pp. 3994–4003 (2016)
41. Pang, J., et al.: Libra R-CNN: Balanced learning for object detection. In: Proc. IEEE Conf. Comput. Vis. Pattern Recognit. (CVPR), pp. 821–830 (2019)
42. Wang, T., et al.: Seesaw loss for long-tailed instance segmentation. In: Proc. IEEE Conf. Comput. Vis. Pattern Recognit. (CVPR), pp. 9695–9704 (2021)
43. Liu, Z., et al.: Balanced meta-softmax for long-tailed visual recognition. In: Adv. Neural Inf. Process. Syst. (NeurIPS) 33 (2019)
44. Zhang, H., et al.: ResLT: Residual learning for long-tailed recognition. IEEE Trans. Pattern Anal. Mach. Intell. (TPAMI) (2021)

Chapter 9
Conclusion and Summary

Abstract This chapter concludes the transformation of object detection and localization, moving from traditional, handcrafted feature models to deep learning systems. A significant driving force behind this change is convolutional neural networks (CNNs) via hierarchical feature extraction, have helped increase accuracy and robustness when compared to previous models. Object detection literature has shifted toward two-stage-detection models, which emphasize accuracy but rely on operations before real-time detection is available, and single-stage models like the YOLO family of detectors, which offer real-time performance—necessary for deployments such as autonomous vehicles and intelligent surveillance applications. This chapter reviews the major areas of object detection applications from autonomous vehicles to medical imaging to intelligent surveillance to demonstrate the far-reaching impact and importance of this detection technology on society. The chapter also covers a handful of important training strategies for training object detection methods that are relevant for developing deep learning models that are both robust and accurate. Some training strategies include data augmentation, use of transfer learning, and a class of specialized loss functions to improve learned feature robustness. While there has been great progress in the object detection field, there are still problems specific to object prefixes that remain, including the ability to detect small, nearby objects, occlusion detection, tradeoffs of computational overhead for edge devices, and dataset bias. Lastly, this chapter identifies numerous achievable research potential areas such as transformer-based modeling architectures, multimodal sensor fusion, 3D scene awareness, federated learning, and explainable AI. With these efforts, we hope to experience more accurate, far-reaching, and ethically leveraged object detection methods, highlighting appealing properties pertinent to the bridge between AI and real-world performance experience.

Object detection and localization represent one of the most remarkable achievements in the history of artificial intelligence and computer vision. They have fundamentally changed machines' abilities to read, analyze, and respond to visual input, representing a progression from detection as simple recognition (response to a pre-established set)

A. Elhanashi and S. Saponara, *Deep Learning for Object Detection and Localization*,
https://doi.org/10.1007/978-981-95-4846-0_9

to recognizing not only what an object is, but more importantly, where it is located. At the heart of object detection is not simply the ability for machines to identify or categorize the object, but to determine the exact location of that object in an image or video stream (e.g. the two-dimensional pixel location of any desired object). The shift from detection as simple recognition to object identification and localization has lead to innumerable real-time, safety-critical, efficiency driven applications in health care, transport, manufacturing, surveillance, agriculture and consumer technologies. The story of object detection is not only a story about object detection algorithms, but also a story about object detection algorithms' integration into society and ultimately, the shifting nature of everyday life because of it. It is helpful to also recognize how far the field of object detection has come and what it grew from—conventional methods associated with traditional computer vision algorithms, object detection integrated past methods based on hand-crafted features like edge detection, histograms of oriented gradients, and color-histograms. Region proposal techniques were introduced to search for likely object locations, followed by classification methods to confirm the presence of an object. These approaches, while pioneering, were highly sensitive to environmental variations, including changes in illumination, scale, perspective, and occlusion. Real-world performance was limited, which prevented widespread deployment in practical applications. The turning point came with the introduction of convolutional neural networks, which automated the process of feature extraction and representation learning. Rather than relying on handcrafted rules, CNNs could hierarchically extract visual patterns directly from raw pixel data. Lower layers captured edges and textures, intermediate layers identified shapes and parts, and deeper layers abstracted semantic representations of full objects. This hierarchical approach brought resilience to variability and enabled much greater accuracy across diverse conditions.

The landscape of object detection began to diverge into two principal families of models: two-stage detectors and single-stage detectors. Two-stage detectors such as R-CNN and its successors introduced region proposal networks that first identified candidate regions before passing them to classification and localization modules. This approach provided very high accuracy, particularly in complex scenes, but suffered from computational heaviness, making it unsuitable for real-time applications. In contrast, single-stage detectors consolidated the functions of localization and classification into a single unified framework, predicting bounding boxes and categories in a single frame. The YOLO and SSD families became emblematic single-stage models responsible for delivering real-time performance in object detection for use in high-stakes, safety-critical settings, such as autonomous driving and robotic deployments. The ongoing development of architectures has attempted to navigate the trade-off between speed and accuracy through hybrid models using a mixture of the two frameworks, and with new innovations such as transformer-based backbones, multi-scale feature pyramids, and attention mechanisms expediting what is possible.

The implications of these models come into focus through their applications. As an example, autonomous driving is one of the most noticeable and socially transforming areas where object detection has become critical. A self-driving car must accurately understand the world around it in real-time: determining the location of pedestrians, cyclists, other vehicles, traffic signs, road boundaries, and even unexpected obstacles. It must operate day and night, through clear skies and fog, in urban intersections and rural highways. The impact of misclassifications could be entirely disastrous; thus, accuracy of predictions is undeniable, and the speed at which those predictions can occur is equally important. Detection systems for pedestrians can be particularly consequential as lives rely on the system's ability to distinguish between fully visible pedestrians and those partially obstructed by vehicles or other environmental factors. Cyclists pose an additional challenge for detection systems due to their small size, speed, and diverse appearance; however, keeping pedestrians and cyclists safe is of the utmost importance. Similarly, it is important to build reliable vehicle detection systems because the same systems that track vehicles also estimate trajectories, plan for lane changes, and predict if a vehicle will stop. Detecting traffic signs (also in poor visibility scenarios) is critical to compliance with the rules of the road. Object detection systems do not only ensure safety; they also increase the convenience of features like automated parking, adaptive cruise control, and navigation. Sensor fusion of radar, lidar, and ultrasonic sensors can also provide an additional dimension of the vehicle's perception system, allowing vehicles to create a more reliable depiction of the environment despite circumstances like poor visibility and/or weather. The use of object detectors and radar has recently enabled the possibility of predictive planning through reinforcement learning, as the vehicle becomes more proactive and anticipates the actions of road users (versus simply reacting after the fact), which is an important step towards reaching true autonomous driving. Object detection has changed surveillance from a labor-intensive, prone-to-error human task into a system that is automated, intelligent, and scalable. Applications in the public safety domain include suspicious conduct, detection of unattended items, and real-time crowd surveillance. Law enforcement agencies use these systems in places such as airports, train stations, stadiums, and downtowns to facilitate situational awareness in order to preemptively and preventively stave off an incident. Traffic monitoring and safety is benefited when multiple intelligent cameras are able to detect various speeders, illegal turns, cumulative congestion patterns in traffic, and accidents, then seize that information for law enforcement consideration, ultimately resolving real safety and efficiency issues. In a retail environment, surveillance systems with object detection provide a plethora of opportunities for stores to analyze customer conduct, respond to the analyzing of customer conduct to improve upon strategic placement of products, and prevent theft by analyzing shopping behaviors over duty of care surveillance practices. In the realm of industrial safety, surveillance system applications often manage compliance with requirements for personal protective equipment, or PPE, management, as well as monitoring unsafe behavior in hazardous working environments. Surveillance systems are even used for healthcare services to help detect falls, monitor compliance with hygiene practices, and assist with patient safety by notifying care team members of emergencies. In the most remote scenarios of application, conservationists are utilizing object detection in wildlife conservation monitoring to track animal movements, discover poachers, or gather ecological data. However, for all of those applications there are challenges such as computational warehouse demands on edge devices; bandwidth for streaming video to a monitoring station; privacy and regulatory concerns such as with the GDPR; and bias of systems in both false positives and false negatives, which complicated application deployments. Models generally need continuous retraining and adaptation to keep up with dynamic environments and people's behaviors.

Medical imaging demonstrates the life-saving potential of object detection. Radiological images (e.g., X-ray, CT, MRI and ultrasound) are key modalities for the diagnosis and treatment of various conditions. Radiologists have historically interpreted radiological images on screens or film in a time consuming manner prone to fatigue and errors. Radiologists can use object detection algorithms to automate parts of this process in radiology practice. For example, these algorithms can identify tumors, lesions, fractures, and other abnormalities in medical images with very high specificity and sensitivity. Convolutional Neural Networks (CNN) based models trained on medical images are capable of learning hierarchical representations of medical images which will detect subtle differences which may not be visible to human inspection of the image. Importantly, these tools are intended to not replace the radiologist but to add to their capabilities by providing a second opinion, decreasing their workload, and increasing the reliability of their assessments. Being able to detect differences sooner facilitates faster intervention and improved outcomes for patients. Transfer learning is an excellent use case for this context because large annotated medical datasets are costly and scarce. Transfer learning allows for pretrained models using a large number of annotated medical images to finetune on a much smaller medical training dataset for reasonable generalization of the model. In addition, few-shot learning methods will be particularly valuable for diseases where count data is limited, since they permit classification from very few observations. Incorporating object detection into medical imaging workflows is an important advance towards more efficient, fair, and accessible healthcare systems.

There are many applications enabled by this advancement due to advances in algorithms and architectures. One important series of object detection models is the YOLO series of models. The YOLOv1 model introduced thinking of detection as a regression problem, which means a single model can predict class probabilities and bounding boxes (location for the prediction) from a single full image. YOLO subsequent versions improved on localization and the ability to detect small objects, such as those you may often see on medical imaging. YOLOv2 incorporated important improvements in batch normalization, anchor boxes, and multi-scale training. YOLOv3 builds on the success of YOLOv2 with the introduction of feature pyramid networks and a stronger backbone of Darknet-53. YOLOv4 examined the latest literature and included many improvements to model structure, including CSPDarknet53, path aggregation networks, and advanced modifications to data augmentation. YOLOv5 gained traction since it was made in a flexible and modular design with implementation in PyTorch, which is an open-source deep learning library for Python. YOLOv6 and YOLOv7 set state-of-the-art results while at the same time optimizing architecture with decoupled heads and extended layer aggregation. Following in the trail of in this path, YOLOv8 created a unified task detection, segmentation, and classification tasks and YOLOv9 through YOLOv11 introduced transformer-based backbones, neural architecture search, quantization-aware training, and federated learning. The pursuit of more lightweight, deployable, and most importantly, accurate predictions across domains is evident through these YOLO versions.

Furthermore, other architectures have their advantages. For example, solitary scale detect (SSD) is able to make use of multi-scale feature extracting techniques in order to maximize object detection across size biasses. On the other hand, EfficientDet is built around the most efficient use of computation and incorporates EfficientNet backbones (with bi-directional feature pyramid networks), which strikes a balance between speed and accuracy. The EfficientDet family of models (from D0 through D7) allows for flexibility within phones, laptops, or powerful servers. Furthermore, attention-based mechanisms utilize predefined modelling around context in order to optimize recognition by cerebrally focusing on certain areas of an image while ignoring noise. Similarly, anchor-free approaches simplify the rectangular bounding box prediction, whereas, both quantizing and pruning a model allow for decreasing the form factor and computation, without significantly affecting accuracy. The list of innovation reflects a larger trend toward designing models that are high-performing in their benchmarking competitions, but also incorporated the practical needs when scaling for deployment (particularly on resource-constrained devices).

Attempting to train these models requires use of thoughtful strategies that combine prepared annotated datasets, augmentation, transfer learning, etc. Training annotated datasets is also arguably our most laborious piece of the ML puzzle, as labeling bounding boxes requires precision from a human. The breadth of diversity and quality of annotations inherently shifts between labels, which varies performance. Dataset augmentation increases variability in our dataset, and also allows our model to be robust, by simulating settings and conditions in the real-world. All models apply both transfer learning and validated transfer learning, with pretrained (or approximated) models being adapted for a performance and computation(s). By freezing lower-layer parameters and modifying higher-layer parameters with lower learning rates, the approach allows the model to adapt to a particular domain without permanently degrading built-in characteristics for generalization. Loss functions are designed to deal with the dual-purposing task of classification as well as localization. Smooth L1 loss functions to handle the bounding box regression and the cross-entropy, or focal loss can handle classification losses, the latter of which is designed to handle class imbalance by focusing on harder samples to classify and has been effective for these cases in imbalanced datasets. Both gradient optimization strategies using SGD with momentum and adaptive optimizers, such as Adam, play a role in how models converge towards a local optimum. Learning rate schedules, such as step decay or cosine annealing work hand in hand with optimization strategies to prevent stuck local minima, and determine the overall performance at final accuracy. Regularization techniques, such as weight decay and dropout, work to curb overfitting while early stopping helps to ensure no resources are wasted on a model that has been trained longer than it should be. Hyperparameter optimization can then be further used to improve model performance. Hyperparameter tuning requires careful adjustment by assessing the optimal learning rate, batch size, weight decay coefficients, and potentially even anchor box dimensions. The choice of batch size can affect the stability in the estimation of gradient convergence, as small batches are better for generalization and larger batches will speed up training times with the risk of falling into sharp local minima. Provided are earlier methodologies to sustain hyperparameter optimization, but there are also several methods that automate and explore the hyperparameter space such as random search, Bayesian optimization, and/or evolutionary algorithms. Tools such as Optuna and Ray Tune provide a scaled framework for hyperparameter tuning and improve the overall iteration cycle of a development cycle.

Domain adaptation refers to addressing when datasets are mismatched, for example, models pretrained on natural images applied to medical imaging, or to datasets applied in autonomous driving. Few-shot and meta-learning frameworks provide the ability to generalize with only a few labeled examples, which helps worse burden large-scale annotation efforts on datasets. These approaches will make object detection feasible, when data is scarce, in certain specialized fields. We evaluate object detection using established metrics, for example, Intersection over Union, precision, recall, mean Average Precision, etc., which allow fair comparisons across the community and therefore, motivate improvements and competition.

Despite these advances in object detection approaches, challenges still exist. Small object detection remains difficult—especially in scenes with a lot of other similar objects occupying the same field of view and low resolution. Occlusion and clutterness demands more sophisticated context modeling and multi-modal schemes. Edge devices and IoT systems establish stringent limits on power, memory and computation; which means, researchers are constantly being pushed to make models more efficient. Dataset bias is detrimental to generalizability and fairness when models are implemented—raising ethical implications when implemented in sensitive contexts like law enforcement and medical settings. Applications involving autonomous vehicles and robots demand ultra-low latency—while maintaining high accuracy—not just from the software, but also from the hardware. Finally, privacy and legal frameworks, such as data protection legislation, exacerbate the complications of deploying systems in situations such as surveillance or medical context.

Object Detection has over many promising directions ahead. By now, Transformer networks have exhibited the ability to catch long-range dependencies, and so will probably be increasingly used in detection pipelines. Multimodal sensor fusion, combining vision with lidar, radar, or other inputs, promises to be a reliable perception in safety-critical domains. 3D object detection and scene understanding will extend the capabilities beyond 2-D bounding boxes. Federated learning approaches will allow collaborative training without necessarily sharing sensitive data and so solve privacy concerns in domains like healthcare and surveillance. End-to-end learning systems encompassing perception, planning, and prediction will simplify decision-making systems, with autonomous vehicles being a clear case for this. Transparency, interpretability, and explainability will be in the center of building trust in AI-driven systems so that their decisions are comprehensible and can be audited and held accountable.

What can be derived from this comprehensive overview is that a discipline that went from a research niche to largely influencing practice in a relatively short time frame. Object detection acts as the bridge between artificial intelligence and the real world, providing the ability for machines to "see", analyze, and act with increasing levels of autonomy. Object detection's success is due to both the actual design of sophisticated algorithms and their careful placement with training, optimizational strategies, and consideration for implementation in the real world. As research progresses, its impact will continue to expand as a disciplinary form of object detection, impacting disciplines and augmenting how humans employ technology, while generating human—machine opportunities. Object detection exemplifies AI that moves beyond abstraction, into systems that engage, in meaningful and safe ways, the realities of the real world.

9.1 Key Takeaways

1. Object detection combines the tasks of identification and localization within machine perception to accurately identify identified items in visual data.
2. Deep learning has sui generis handcrafted features with hierarchical representations able to generalize across rotations and translations in the real world.
3. Two-stage detectors trade for accuracy, and single-stage detectors such as YOLO and SSD trade-off for real-time implementations.
4. Applications range from automotive, surveillance, medicine, agriculture, robotics, augmented reality, demonstrating flexibility in the observed phenomenon of object detection.
5. The latest developments of object motion detection with software-based approach include YOLOv11 and EfficientDet to maintain trade-offs for current real-time accuracy and efficiency for implementations on edge devices and IoTs.
6. Training depends on annotated datasets, augmentation, transfer learning, and the creation of a loss function and optimization strategy.
7. Hyperparameter tuning, domain adaptation, and few-shot learning are important aspects of training a model to fit a more specialized domain or working with limited data.
8. Some of the unsolved issues remain small-object detection, occlusions, efficient computation, bias in dataset and ultra-low latency.
9. Standardized metrics (i.e. intersection over union, precision, recall and mean average precision) create standardized measures that help create relevant competition amongst improving with benchmark measures.
10. The future still includes variants of transformer-based detection, fusion of multimodal sensor data, three-dimensional scene understanding, federated learning and interpretable ai to help build even more powerful, reliable and ethically developed systems.

Glossary

Object Detection Finding and identifying the objects in an image.

Localization The process of determining the precise spatial position of a detected object within an image.

Bounding Box A rectangle drawn around a detected object.

Convolutional Neural Network (CNN) A deep learning model designed for image processing.

Two-Stage Detector A high accuracy, slower detector that first process the regions, then classify them.

Single-Stage Detector A faster detector that predicts an object in one step.

Regional proposal Network (RPN) A Network that suggests potentional objects location.

Anchor Box A predefined box that is used as a template for prediction.

Intersection Over Union (IOU) A meteric measuring the overlap between a predicted box and truth box.

Non-Maximum Suppression (NMS) A A technique to remove duplicate bounding box.

Feature Pyramid Network (FPN) An architecture that helps detects object at different scale.

Mean Average Precision (mAP) The main accuracy metric for object detection model.

Object Tracking A computer vision task that follows a specific object or (multiple objects) through consecutive frames in a video sequence.

Transfer Learning Using a pre-trained model to start a new task.

Real-Time Processing Analyzing data instantly for video applications.

Data Annotation Labeling data to train AI model.

Semantic Segmentation A computer vision technique where every pixel in an image assigned a single class label.

Data Augmentation Creating new training images by altering existing ones.

Normalization Scaling image pixels to a standard range.

Synthetic Data Artificially generated data.

A. Elhanashi and S. Saponara, *Deep Learning for Object Detection and Localization*, https://doi.org/10.1007/978-981-95-4846-0

Class Imbalance Where some data classes have far more examples than others.

You Look Only Once (YOLO) A family of single-stage object detection models that are known for their speed and good accuracy.

SSD (Single Shot Multibox Detector) A single-stage detector that relies on multiple feature maps to detect objects in the image at multiple scales.

EfficientDet A family of object detectors that are optimized through a compound scaling approach to weigh the model size, speed, and performance efficiently.

CenterNet An anchor-free, single-stage detector which treats an object (everything) as a single point—its center—and predicts all the properties from the center point.

Regional Convolutional Neural Network (R-CNN) A two-stage detection process that proposes regions of interest using selective search and feeds these into a CNN for classification.

Faster Regional Convolutional Neural Network (Faster R-CNN) A two-stage detector that introduced the "Region Proposal Network" (RPN) which is able to integrate the proposal step within the network allowing for end-to-end training.

Confidence Score A type of probability (with a range of 0 to 1) that indicates how certain a model is that a detected bounding box contains an object.

DETR—Detection Transformer A transformer-based object detection model that frames detection as a set prediction problem, thereby removing the need for hand-designed components such as anchor boxes or NMS.

End-to-End Trainable System AA model that is trained as a single system across an input to output pipeline, without requiring separate stages or sub components.

Ground Truth The human-annotated data and definitive reference for object location, used for training, validation, and evaluation of detection models.

Instance Segmentation An anchor-free, single-stage detector which treats an object (everything) as a single point—its center—and predicts all the properties from the center point.

Autonomous Driving Application of object detection and localization to provide self-driving vehicles the ability to understand their environment by recognizing objects.

Surveillance Systems Security-based applications that leverage object detection to monitor public and private environments in real-time, automatically and for public safety and traffic management.

Medical Imaging Application of object detection in the healthcare domain to recognize anatomical structures automatically, and be able to localize abnormalities such as tumors in various imaging modalities (e.g. X-ray, MRI, CT).

Robotics Application of object detection as part of a robot system to detect and interact with objects in their environment for various applications.

Edge Computing The processing of data and/or model inference on a local device, rather than in the cloud, for the benefit of lower latency and using less bandwidth. For example, in an Internet of Things (IoT) application, or in some medical applications.

Federated Learning A machine learning approach that allows for model training to occur across a number of decentralized devices, without sharing raw data, thus particularly useful for approaches that aim to protect privacy (e.g. in medical imaging, healthcare, etc.).

Explainable AI (XAI) Approaches to explain or its decision making approach, to humans; especially why a decision is made such as object detection. Key for trust especially in medical or safety-critical applications.

Deep Learning Framework A software library or toolset, such as TensorFlow or PyTorch, that gives the elements to create, train, and validate deep learning models, including those used for object detection.

TensorFlow An open-source deep learning framework created by Google, featuring production-ready deployment tools and a full ecosystem with TensorFlow Lite for mobile devices.

PyTorch is an open-source deep learning framework that is widely used for research and prototyping because of its dynamic computation graph and Pythonic and intuitive.

Model Compression Methods such as pruning and quantization that decrease the size and processing requirements of a neural network model to be able to efficiently deploy them on resource-constrained devices.

PyTorch is an open-source deep learning framework that is widely used for research and prototyping because of its dynamic computation graph and Pythonic and intuitive.

Quantization It is the process of approximating a neural network that uses floating-point numbers by a neural network of low bit width numbers.

Edge Deployment The process of performing deep learning algorithms and models directly on or in close proximity to the data-generating devices, such as "edge devices," constitutes edge AI, of which there are also hybrids, again developing models and algorithms in the cloud and transferring them.

Cloud Deployment Refers to the process of designing, deploying, and running software and workloads in cloud computing environments, including public, private, or hybrid clouds, in order to make applications available to end users in an on-demand mode.

Hyperparameter Optimization The methodical process of exploring hyperparameters to find the optimal hyperparameters (e.g., learning rate, batch size) that maximizes model performance.

Learning Rate A hyperparameter that defines the step size of each update of the model as it's trained; the learning rate determines the amount of change to the model in response to the estimated error each time the model weights are updated.

Loss Function A function that tells how far apart the model predictions are from ground truth, the loss function helps define how the model is optimized during training.

Overfitting An error in modeling that occurs when a model learns, too well, the training data (including noise and outliers), resulting in poor performance on new data that was not part of the training set.

Background Clutter The occurrence of intricate or noisy backgrounds that may be mistaken for objects, resulting in positive detections.

Intra-Class Variability The variation of visual appearances (e.g., shapes, colors, textures) that can occur within an object class that can impact the consistency in detection.

Index

A. Elhanashi and S. Saponara, *Deep Learning for Object Detection and Localization*,
https://doi.org/10.1007/978-981-95-4846-0

MIX
Papier aus verantwortungsvollen Quellen
Paper from responsible sources
FSC® C105338

If you have any concerns about our products,
you can contact us on
ProductSafety@springernature.com

In case Publisher is established outside the EU,
the EU authorized representative is:
Springer Nature Customer Service Center GmbH
Europaplatz 3, 69115 Heidelberg, Germany

Printed by Libri Plureos GmbH
in Hamburg, Germany